# 100 scientific discoveries
## that changed the world

Cover: Discoveries such as the Internet (background), microorganisms, cell phones, cars, and genetic engineering (inset, left to right) have shaped humankind's worldview and forever altered our future. Below: Combining two important scientific discoveries, telescopes at the Starfire Optical Range in New Mexico beam lasers into the night sky that collide in the atmosphere to create an artificial star.

# 100
# scientific
# discoveries
## that changed the world

NATIONAL GEOGRAPHIC

WASHINGTON, D.C.

# Contents

(1-25) COMPUTING, COMMUNICATIONS, AND THE NANOWORLD
From the abacus to cloud computing, the manipulation of ones and zeros has made the world a faster, more interconnected place.

(26-50) HEALTH AND MEDICINE
If yesterday's medical pioneers were disease detectives, today's doctors are creators whose material is the genetic code itself.

(51-77) PHYSICS AND ENGINEERING
Fundamental questions about what moves the world have led to practical advances such as automobiles as well as mind-boggling visions of a multidimensional universe.

(78-100) EARTH SCIENCE AND ASTRONOMY
Discoveries about the age and ecology of our planet have paced hand in hand with the realization that Earth is just one world among many in an expanding universe.

Albert Einstein's (left) contributions to science range from discoveries involving relativity to the foundations for laser technology.
A flower beetle is fitted with a micro-electrical-mechanical system (above).

# Foreword

The original big ideas came from innovative men and women whose names are long lost. No monuments commemorate the inventors of the bowl, the dugout canoe, or the wheel, nor those who first planted crops, smelted copper, or etched marks into wet clay to inaugurate writing. Yet their legacies are all around us, in the foundations of the modern world.

With the advent of writing, big ideas came to be regarded as the providence of big thinkers. These intellectuals, as they were called, contributed valuable insights but seldom discovered or invented anything. Instead they analyzed and rearranged the relatively few facts that were known then, like jailhouse cardsharps forever shuffling the same deck of cards.

Science and technology did not so much build on the intellectual tradition as react against it, as practitioners returned to the habits of hands-on tinkering that characterized prehistoric innovation. Pioneering scientists like Galileo, Gilbert, Harvey, and Newton had little use for scholarly analysis of venerable opinions. They were more apt to agree with Francis Bacon, the great 17th-century prophet of science, who likened his Cambridge professors to "becalmed ships; they never move but by the wind of other men's breath."

The scientists preferred to find new facts, such as how gravity and magnetism work, how blood circulates through the human body, and how planets orbit the sun. Their points of reference came less from reading old books than from experimentation and observation—what Galileo called reading "the book of nature."

The result of their campaign was an unprecedented improvement in the lives of people around the world. Prior to the scientific and technological revolutions, the average human being was illiterate, earned a few hundred dollars a year, and was unlikely to survive to see age 30. Today, over 80 percent of all adults are literate, the global median annual income exceeds $7,000, and life expectancy at birth is approaching age 70. All this happened so quickly, measured against the long shadows of our history and prehistory, that many people don't yet realize that it has happened.

I keep on my desk a Neanderthal hand ax, chipped from a piece of obsidian some 34,000 years ago. It's a good ax; pick it up and you immediately start imagining all sorts of things you could do with it, from cutting meat to defending yourself to making another ax. That spirit, one of learning and being inspired through direct physical interrogation of nature, is the real impetus behind science and technology alike.

Bacon, writing at the dawn of modern science, argued that experimenters "are like the ant; they only collect and use," whereas logicians "resemble spiders, who make cobwebs out of their own substance.

"But the bee takes a middle course," Bacon wrote. "It gathers its material from the flowers of the garden and of the field, but transforms and digests it by a power of its own."

Time proved Bacon right. Scientists today are so immersed in technology, and technologists in science, that it can be difficult to trace where one ends and the other begins. This messy process satisfies the neat prescriptions of neither scholars nor priests, but its results speak for themselves: More facts are now discovered in a decade than were once acquired in a century.

Were Bacon alive today he might compare global science and technology to fields of wildflowers fertilized by bees—astonishing in their variety, yet each part testifying to the nature of the whole. The volume you are holding is a way into that excitement and splendor. Welcome, in short, to a beehive of a book.

**Timothy Ferris, author of** *Coming of Age in the Milky Way*

# About This Issue

## BUILDING ON THE PAST

The scientific discoveries profiled in this issue run from 1 to 100 and are organized by field. In each section they follow a reverse chronology, so that the most recent discoveries build on the work that preceded them.

### Discovery Pages

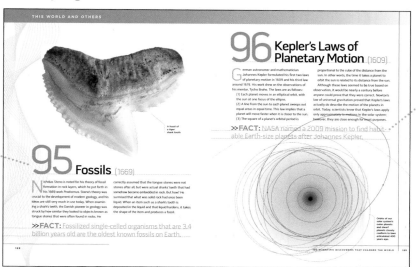

On these pages, each numbered finding is accompanied by its date of discovery.

Additional facts highlight extra information or connections with other discoveries.

### Breakthrough Pages

Breakthrough pages are devoted to 12 ideas that reflect the cutting-edge thinking of today.

# 01

# THE POW INFORMA

Electronic, wireless information and communication technologies are ubiquitous in today's world. However, just a generation ago, cell phones, the Internet, and personal computers were not parts of everyday life—although each had been invented. What were the key ideas behind these world-altering technologies? How did they evolve from earlier discoveries? What does the future of information technology look like?

Ideas never develop in isolation. Rather, they build on one another. Since the 17th century a sophisticated understanding of binary numbers has played a central role in both computer and telecommunications technologies. Computers are logic machines that operate on Boolean logic, which was described in the 19th century. The transistor was invented in the 1940s, and the integrated circuit—the basis of today's microprocessor—came along a decade later, making the personal computer a possibility, although not a reality until the 1970s.

# ER OF
# TION

In the late 1940s—about a decade after the formulation of Turing theory—mathematician Claude Shannon developed a theory of communication now known as information theory, which forms the foundation of nearly all modern electronic communications. Making its way into peoples' lives today is augmented reality—a union of computer and communications technologies that exploits the global positioning system, wireless communication, and ever smaller yet increasingly powerful computational devices.

Where does the story go from there? Smaller, faster, cheaper—and the world is increasingly connected by way of the infrastructure known as the Internet. Now the Internet is being harnessed to create cloud-computing applications, and, with some success, computer programs are being created with not only the ability to learn, but also the ability to manipulate matter on the scale of single atoms.

Computer artwork depicts a nanotube bent into a ring. Nanolithography is a method used to create circuit boards.

# 1 Nanolithography (1999)

Today's cell phones, computers, and GPS systems would not be as compact as they are without the technique known as nanolithography, one branch of the revolutionary science of nanotechnology.

Nanolithography is a way of manipulating matter on the scale of individual atoms in order to create circuit boards for a variety of electronic devices. Through the use of an atomic-force microscope, atom-size nanomaterials such as nanocrystals, nanolayers, and nanotubes are arranged into structures. Dip pen nanotechnology, developed in 1999 by Chad Mirkin of Northwestern University, has allowed circuit boards to become much smaller. This, in turn, has led to the development of computers so tiny that they could be used in other nanoscale technologies, such as programmable matter.

>> **FACT:** Nanolithography has its foundations in the invention of the microscope in 1590.

# 2 Carbon Nanotubes (1991)

In 1991 Japanese physicist Sumio Iijima discovered carbon nanotubes, considered one of the most important discoveries in the history of physics. Nanotubes can be constructed by an arc evaporation method, in which a 50-amp current is passed between two graphite electrodes in helium. The results are nanotubes that measure 3 to 30 nanometers in diameter.

One of the amazing properties of carbon nanotubes is their strength. Their resistance to stress is five times that of steel, and their tensile strength is up to 50 times that of steel. Carbon nanotubes can also be used as semiconductors. Some nanotubes' conductivity is greater than that of copper, for example.

Scientists and engineers are looking for ways to use nanotubes in the construction industry, as well as in aerospace applications. Today, flat panel displays and some microscopes and sensing devices incorporate carbon nanotubes. In the future many everyday items—from homes to computer chips to car batteries—might be made of carbon nanotubes.

Carbon nanotubes are made up of rolled sheets of carbon atoms. The tubes may find use in tiny electrical components.

# 3 World Wide Web (1990)

The World Wide Web has changed lives forever by linking millions of computers throughout the world, thus making all the information they contain available to everyone. Although the terms *World Wide Web* and *Internet* are often used interchangeably, they are not the same thing. The Internet is the system that links the network of information that is the Web. It is possible to have the Internet without the Web, but the Web cannot exist without the Internet.

The Internet began in 1962 as ARPANET, a network of two computers conceived of by the U.S. military. That network grew to more than a million computers by 1992. In 1990 British computer scientist Tim Berners-Lee invented the Web—that is, he created the first Web browser and Web pages, which could be accessed via the Internet. Web pages incorporate hypertext—text displayed on a monitor that contains hyperlinks to other documents—a concept that dates to the 1960s. And the rest, as the saying goes, is history.

The most profound way in which the Web has affected economic and social life is the ease of communication it created. The Web is a giant global community. People in Canada, South Africa, and Australia can share ideas with one another or engage in a debate about current events. A woman in Moscow can purchase a collectible item from a man in Brazil who has the exact piece she is looking for.

A global "map" of the Internet shows the extent of the network's

# 4 Buckyball (1985)

As early as 1965, scientists predicted the theoretical existence of a molecule made of multiple carbon atoms and shaped as a sphere or a cylinder. But it was not until 20 years later that Richard Smalley and Robert Curl, both professors at Rice University in Houston, Texas, and Harry Kroto, a professor at the University of Sussex, England, discovered one. By focusing lasers on graphite rods, they generated molecules shaped symmetrically, somewhat like geometrically regular cages. Seeing in these remarkable molecules a similarity to the geodesic dome—a construction made of a lattice of triangles and designed by visionary architect R. Buckminster Fuller—they named the newfound molecule buckminsterfullerene, or fullerene—buckyballs for short. The three researchers received the Nobel Prize in chemistry in 1996.

The discovery of fullerene has scientific and technological implications, and the molecule has many practical uses. Analysis of buckyballs has advanced understanding of the behavior and manipulability of sheet metals. Excellent conductors of heat and electricity, fullerene materials may replace silicon-based devices in computers, cell phones, and similar electronics. The material also exhibits incredible tensile strength, thus promising new possibilities in architecture, engineering, and aircraft design.

In 1991, Japanese researcher Sumio Iijima's discovery of an oblong version of the buckyball, ultimately called the nanotube, spurred the nanotechnology revolution of the early 21st century.

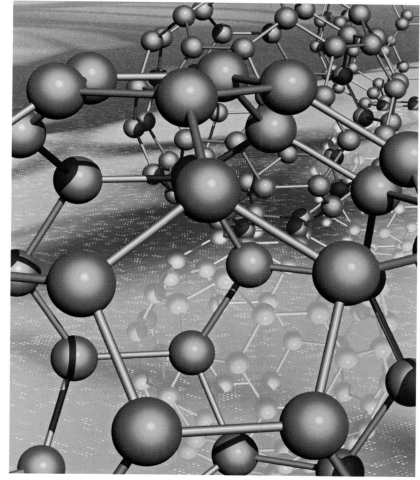

A fullerene (detail, above, and above left) is a lattice-like molecule composed of carbon. Named for architect R. Buckminster Fuller, it is called a buckyball for short.

# 5 Global Positioning System (1978)

Located 12,000 miles above Earth and traveling at a speed of 7,000 miles per hour, the satellite-based navigation system was launched in 1978 by the U.S. Department of Defense for military purposes. Shortly afterward, the manufacturers of GPS equipment recognized its mass-market potential and clamored for its use as a civilian application. The Department of Defense complied with their request in the 1980s.

GPS, which is extremely accurate, works on the principle of triangulation, by which the position of a given object (a person in a car, for example) is determined by the distance of the object from four satellites in the network with which a device is communicating. The GPS network is composed of 24 satellites, with 3 on standby in case of failure, and each satellite makes two complete revolutions around the planet every day.

So what does the future of GPS look like? To be sure, the network will continue to become more accurate and to provide increasingly fine-grained information on a given location. Drivers still need to take their eyes off the road to look at the GPS device, but that may change. A company called Making Virtual Solid is working on a solution called Virtual Cable, which is designed to be built in to a car's windshield. A red line that appears to follow the landscape guides the driver to his or her destination.

>> **FACT**: The 24 satellites in the GPS network run on solar energy.

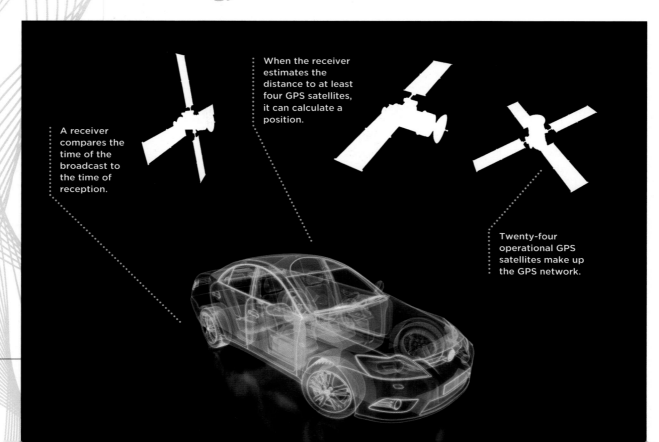

A receiver compares the time of the broadcast to the time of reception.

When the receiver estimates the distance to at least four GPS satellites, it can calculate a position.

Twenty-four operational GPS satellites make up the GPS network.

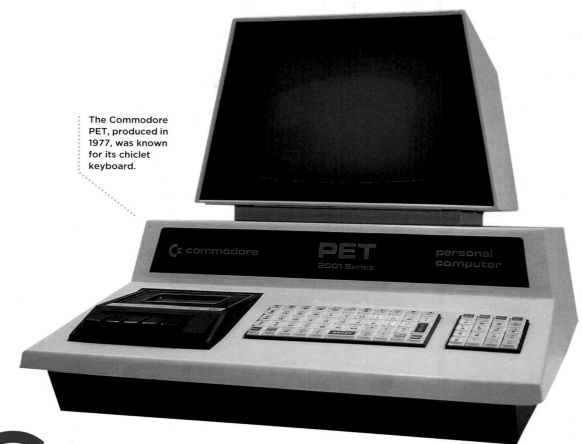

The Commodore PET, produced in 1977, was known for its chiclet keyboard.

# 6 Personal Computer (1977)

According to the Nielsen Company, over 80 percent of U.S. households had a home computer in 2008, and of those, about 90 percent had Internet access. Whether it be an early model—like the Commodore PET (very popular in schools in the late 1970s), the Apple II (one of the first highly successful mass-produced microcomputers), and the IBM PC (designed to supplant the first two devices in homes)—or the latest and greatest in machines, all PCs have the same basic components. A computer is made up of a motherboard, a processor, a central processing unit, memory, drives, a fan, and cables. Attached to the computer are its peripherals: the mouse, keyboard, monitor, speakers, printer, scanner, and so on. These components work together to run software: the operating system and additional programs, such as a word-processing program, money-management software, or photo-editing software.

With the personal computer, computing technology became available to the general public. Computers were no longer large, highly expensive pieces of equipment that only major corporations or government agencies could afford or only computer scientists could operate. This significant development ultimately birthed new industries, changed how people communicate, and irrevocably altered their work and personal lives.

# 7 Augmented Reality

Unlike virtual reality, which is based in a computer-generated environment, augmented reality is designed to enhance the real world by superimposing audio, visual, and other elements on your senses.

Boeing researcher Tom Caudell first coined the term *augmented reality* in 1990 to describe the digital display used by airline electricians, which combines virtual graphics with physical reality. But the concept is even older than that. The 1988 movie *Who Framed Roger Rabbit* is a good example of the technology—and, originating before that, a more simplistic version is demonstrated in the yellow arrows that announcers use on televised football games when analyzing a play.

Augmented reality is currently driving the development of consumer electronics because of its usefulness in smart-phone applications. An app called Layar uses a cell phone's camera and GPS capabilities to gather information about the surrounding area. Layar then shows information about certain nearby sites, such as restaurants or movie theaters, and overlays this information on the phone's screen. Point the phone at a building, and Layar reports if any companies in that building are hiring or locates the building's history in the online encyclopedia Wikipedia.

There are some limitations to augmented reality technology as it currently exists: The GPS system has a range of only about 30 feet, the screens of cell phones are small, and there are understandable concerns about privacy, especially as the technology affects more and more aspects of our lives. Even so, the future of this technology is bright—with obvious, soon-to-be-tapped potential for gaming, education, security, medicine, business, and other areas.

• **5** The **GLOBAL POSITIONING SYSTEM** is essential to augmented reality, which depends on GPS to pinpoint a person's location.

• **10** SMART PHONES are the devices on which most augmented reality technologies are currently implemented.

### Learn More

Like captions on a page, augmented reality can provide background and explanations, such as a biography of architect Frank Gehry when visiting his signature buildings.

### Orient and Navigate

Global positioning systems allow geographical tags that map a person's location at a moment's notice and then plot surrounding sites and pathways.

### Make Choices

Consider options and make choices virtually. Here, galleries are represented by images of museum holdings, such as Richard Serra's "La Materia del Tiempo" (shown).

Use a smart phone to view cues connected to an actual place, such as the Guggenheim Museum Bilbao, and you can layer on information.

HROUGH

# 8 Public Key Cryptography (1976)

nvented in 1976 by Stanford University professor Martin Hellman and graduate student Whitfield Diffie, public key cryptography is a security technology that enables people using an unsecured network (like the Internet) to transmit private data such as bank account numbers securely. Here is how it works: A certificate authority simultaneously creates a public "key" and a private "key." Much as a key unlocks a door, these keys are particular values that unlock encrypted data. The private key is given only to the person requesting it, and the public key is made publicly available. The private key is used to decrypt information that has been encrypted with the public key. It is complex, but the end result is what matters: Whether buying a coveted collectible on eBay, paying bills, or renewing a driver's license online, public key cryptography protects personal data.

Public key cryptography ensures that confidential data transmitted over the Internet remains protected.

**>> FACT:** Public key cryptography has closed many security loopholes in wireless transmission.

# 9 Molecular Electronics (1974)

As its name implies, molecular electronics refers to the use of molecular components to build electronic devices. Since chemists Mark Ratner and Ari Aviram created the first molecular electronic device in 1974—a rectifier, which converts alternating current to direct current—scientists have continued to advance both their understanding of the science and its potential applications.

Many researchers are working to replace semiconductors in all of their applications with molecular electronic switches. Some companies are poised to deliver such switches to computer and electronic-device manufacturers. One example is a Huntsville, Alabama–based company called CALMEC, which has created a molecular-size switch. This device can be used in electronic semiconductors, thus enabling electronic technology to be miniaturized even more than it is today.

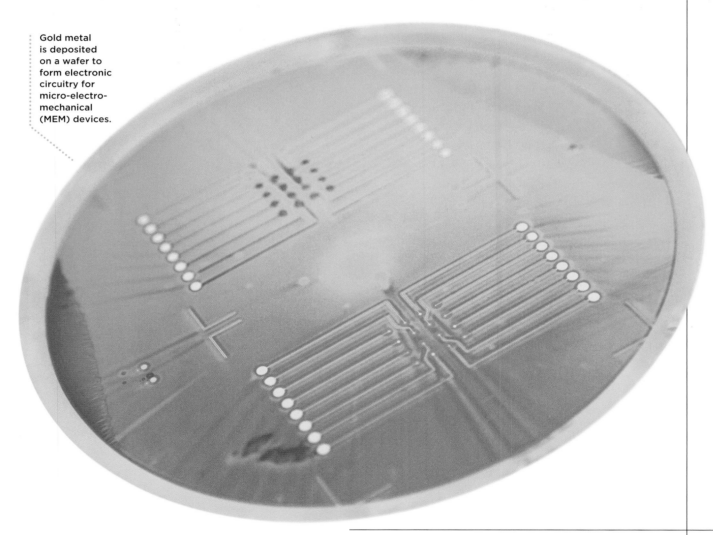

Gold metal is deposited on a wafer to form electronic circuitry for micro-electro-mechanical (MEM) devices.

# 10 Cell Phones (1973)

It is hard to say what is more ubiquitous in today's world—the personal computer or the cell phone, which was invented in 1973 by Martin Cooper when he was the director of research and development at Motorola.

A cell phone is actually a radio, albeit a highly sophisticated one. It is a full-duplex device, which means that two different frequencies are used for talking and listening. The communication occurs on channels, of which the average cell phone contains more than 1,650.

In a typical cell phone network, a carrier, which provides the cell phone service, is assigned 800 frequencies, which are divided into hexagonal units called cells. Each cell contains about ten square miles. Each cell has its own base station and tower. Both cell phones and cell towers have low-power transmitters in them. The phone and the tower use a special frequency to communicate with each other. (If this frequency cannot be found, an "Out of range" or "No service" message is displayed on the phone's screen). As a caller uses the phone and moves from one cell to another, the frequency is passed from one cell to the next. The carrier maintains the frequency needed to communicate with the person on the other end and continually monitors the signal strength. If the caller moves from one carrier's network to that of another, the call will not be dropped, but the caller's jaw may drop when he sees the roaming charges on his bill.

The first cell phones were bulky and had antennas (left); today's smart phones (below) are smaller and far more powerful.

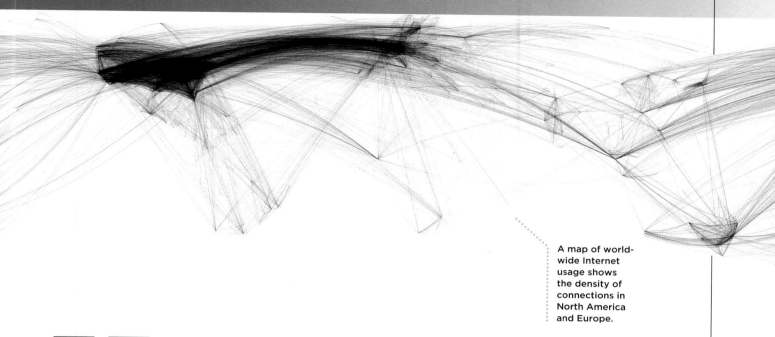

A map of world-wide Internet usage shows the density of connections in North America and Europe.

# 11 Internet (1969)

Thanks to the Internet, more aspects of life are moving online than ever before. What began as a collaboration among academia, government, and industry in the late 1960s and early 1970s has evolved into a vast information infrastructure.

The Internet works because of a few technologies. The first is packet switching, whereby data are contained in specially formatted units, or packets, that are routed from source to destination via network switches and routers. Each packet contains address information that identifies the sender and the recipient. Using these addresses, network switches and routers determine how best to transfer the packet between points on the path to its destination.

The Internet is also based on a key concept known as open architecture networking. In a sense, this is what makes the Internet the Internet. With this concept, different providers can use any individual network technology they want, and the networks work together through an internetworking architecture. Thus these networks act as peers for one another and offer seamless end-to-end service.

A third important technology is transmission-control protocol/Internet protocol, or TCP/IP. This is what makes an open architecture network possible. Think of it as the basic communication language of the Internet. TCP assembles a message or file into smaller packets that are transmitted over the Internet, and IP reads the address part of each packet so that it gets to the right destination. Each gateway computer on the network checks this address to determine where to forward the message.

>>**FACT:** It is estimated that over two billion people worldwide accessed the Internet in 2011.

# 12 Nanotechnology (1959)

By late 1959 the goal of creating smaller and smaller devices was on the minds of scientists and researchers, and they had made some progress. For example, inventors had developed motors that were about the size of a fingertip.

Richard Feynman, a California Institute of Technology physics professor, envisioned far greater advances. On the evening of December 29, 1959, he gave his now famous speech on nanotechnology at an event for the American Physical Society. In his speech, titled "There's Plenty of Room at the Bottom," he described the ability to write the entire *Encyclopedia Britannica* on the head of a pin by using atom-size tools or machines.

Feynman's vision of nanotechnology considered its many practical applications. He based his then revolutionary ideas on the fact that each living cell of an organism contains all of the genetic information needed to create that organism. This was his evidence that storing vast amounts of data in minute objects was possible.

Fire-resistant glass contains a layer of silica nanoparticles that help panes withstand temperatures of up to 1800°F for more than two hours.

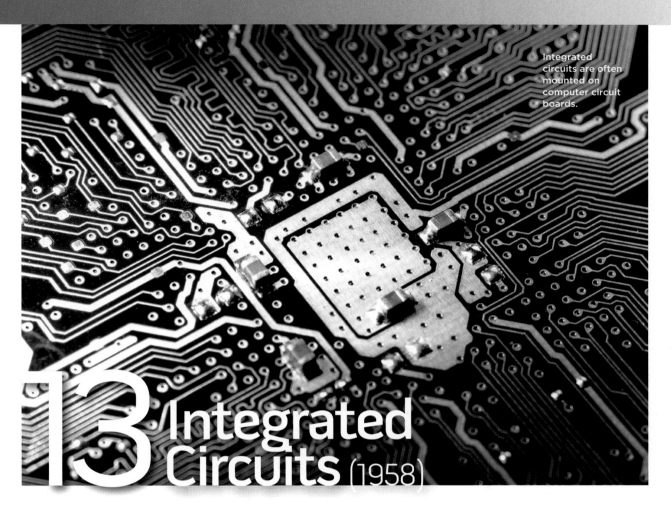

Integrated circuits are often mounted on computer circuit boards.

# 13 Integrated Circuits (1958)

ntegrated circuits (ICs) are found in nearly every electronic device used today, from cell phones to television sets. A complex electronic circuit, each IC contains a diode, a transistor, a resistor, and a capacitor. These components work together to regulate the flow of electricity through a device. But integrated circuits have disadvantages: All connections must remain intact or the device will not work, and speed is definitely a factor. If the components of the IC are too large or the wires connecting the components are too long, for example, the device is slow and ineffective.

In 1958 Americans Jack Kilby and Robert Noyce separately solved this problem by using the same material to construct both the integrated circuit and the chip it sat on. Wires and components no longer had to be assembled manually. The circuits could be made smaller, and the manufacturing process could be automated. (To demonstrate just how small these circuits can be, consider this: The original IC had only one transistor, three resistors, and one capacitor, and it was the size of an adult's pinkie finger. Today, an IC smaller than a penny can hold 125 million transistors.)

In 1961 the first commercially available integrated circuits were introduced, and computer manufacturers immediately saw the advantage they offered. In 1968 Noyce founded Intel, the company that introduced the microprocessor, which took the IC one step further by placing a computer's central processing unit, memory, and input and output controls on one small chip.

>> **FACT:** The invention of the semiconductor in 1911 paved the way for the development of integrated circuits.

# Cloud Computing

Visit social media websites—such as Facebook, Twitter, Flickr, or National Geographic's Your Shot (shown here)—and you're entering the cloud.

In the past, computing relied on a physical infrastructure: routers, data pipes, hardware, and servers. These items have not gone away—nor are they likely to disappear altogether—but the process of delivering resources and services is moving to a model whereby the Internet is used to store the necessary applications.

An immediate benefit of this model is lower cost. For example, companies no longer have to buy individual software licenses for every employee. With cloud computing, a single application gives multiple users remote access to the software. Web-based email, such as Google's Gmail, is an example of cloud computing.

To understand the concept of cloud computing, it helps to think in terms of layers. The front-end layers are what users see and interact with—a Facebook account, for example. The back end consists of the hardware and software architecture that runs the user interface on the front end. Because the computers are set up in a network, the applications can take advantage of all the combined computing power as if they were running on one particular machine.

While there are advantages to this model, it is not without drawbacks. Privacy and security are two of the biggest concerns. After all, a company is allowing important, potentially sensitive data to reside on the Internet, where, in theory, anyone could access it. The companies that provide cloud-computing services, however, are highly motivated to guarantee privacy and security—their reputations are at stake. An authentication system that employs user names and passwords or other types of authorization helps to ensure privacy.

• **3** Cloud-computing applications, by definition, reside in the **WORLD WIDE WEB.**
• **6** The **PERSONAL COMPUTER** is the means by which most people will access the cloud.
• **8 PUBLIC KEY CRYPTOGRAPHY** enables the encrypting of electronically transmitted private information.

• **11** The **INTERNET** is the foundation of cloud computing and allows users to run applications on computers other than their own.
• **19** A **LARGE-SCALE ELECTRICAL SUPPLY NETWORK** is necessary for the use of a cloud-computing infrastructure.

Cloud computing allows participants to access digital materials without the data's residing in the hardware of their computers.

# 15 Information Theory (1948)

Information theory has its origins in a 1948 paper by American mathematical engineer Claude Shannon: "The Mathematical Theory of Communication." This theory allows the information in a message to be quantified, usually as bits of data that represent one of two states: on or off. It also dictates how to encode and transmit information in the presence of "noise," which can corrupt a message en route.

At the heart of Shannon's theory is the concept of uncertainty. The more uncertainty there is with regard to what the signal is, the more bits of information are required to make sure the essential information is transmitted. Shannon called this uncertainty-based measure of information entropy. He mathematically proved that a signal could be encoded—reduced to its simplest form, thus eliminating interference or noise—to transmit a clear message, and only the message. Although there is always a possibility of error in transmission, the application of information theory vastly minimizes that possibility.

Via coding theory, an important offshoot of information theory, engineers study the properties of codes for the purpose of designing efficient, reliable data by removing redundancy and errors in the transmitted data. It has two primary characteristics. Source encoding is an attempt to compress the data from a source in order to transmit the data more efficiently (if you have ever "zipped" a file to send it to someone, you have seen source encoding in action). Channel encoding adds extra data bits to make the transmission of data more resistant to disturbances present on the transmission channel.

A music-mixing console electronically combines, routes, and changes the dynamics of audio signals.

>> **FACT:** Information theory developed out of Shannon's desire to clear up the "noise" in telephone conversations.

# 16 Transistor (1947)

A transistor is a type of semiconductor, without which modern electronic devices—including computers—could not function. Although there are several different types of transistors, all contain a solid piece of semiconductor material, with at least three terminals that can be connected to an external circuit. This technology transfers current across a material that normally has high resistance (in other words, a resistor); hence, it is a transfer resistor, shortened to transistor.

Prior to the introduction of the transistor, computers operated by way of vacuum tubes, which were bulky and expensive to produce. The more powerful computers contained thousands of them, which is why early computers filled entire rooms. In 1947 American physicists John Bardeen and Walter Brattain at Bell Labs observed that when electrical contacts were applied to a crystal of germanium, the power generated was greater than the power used. American William Shockley, also a physicist, saw the potential in this, and over the next few months the team worked to expand their knowledge of semiconductors. In 1956, the three men won the Nobel Prize in physics for inventing the transistor.

So why is the transistor so important to modern electronics? Among other benefits, it can be mass-produced using a highly automated process for a relatively low cost. In addition, transistors can be produced singly or, more commonly, packaged in integrated circuits with other components to produce complete electronic circuits. And transistors are versatile, which is why they are used in practically every electronic device known today.

The transistor was invented in 1947 to replace the bulky and expensive vacuum tube.

Geostationary satellites orbit Earth.

# 17 Communication Satellites (1945)

Although most people know him as a prolific science-fiction author, Arthur C. Clarke made a significant contribution to communications technology. In October 1945 he outlined his concept of geostationary communication satellites in a paper titled "Extra-Terrestrial Relays—Can Rocket Stations Give Worldwide Radio Coverage?" Although Clarke was not the first to come up with the theory, he was the first to popularize it.

The term *geostationary* refers to the position of a satellite's orbit around Earth. The orbit of a geosynchronous satellite repeats regularly over specific points. When that regular orbit lies over the Equator and is circular, it is called geostationary.

The advantages of geostationary satellites are many. Receiving and transmitting antennas on the ground do not need to track the satellites, since they do not waver in their orbits. Nontracking antennas are cheaper than tracking antennas, so the costs of operating such a system are reduced. The disadvantages: Because the satellites are so high, radio signals take a little longer to be received and transmitted, resulting in a small but significant signal delay. In addition, these satellites have incomplete geographic coverage since ground stations at higher than roughly 60° latitude have difficulty reliably receiving signals at lower elevations.

Regardless of the disadvantages, there is no denying that communication satellites have revolutionized such areas as global communications, television broadcasting, and weather forecasting, and they have important defense and intelligence applications.

# 18 Radio Waves (1888)

Scottish physicist James Clerk Maxwell was one of the first scientists to speculate about the nature of electromagnetism. The equations he formulated describe the behavior of electric and magnetic fields, as well as their interactions with matter. Maxwell theorized that electric and magnetic fields travel through empty space, in the form of waves, at a constant velocity. He also proposed that light waves and radio waves are two forms of electromagnetic radiation.

In 1888, German physicist Heinrich Hertz became the first person to demonstrate Maxwell's theory satisfactorily when he proved the existence of radio waves. Hertz did this by building a device that could detect very high frequency and ultrahigh frequency radio waves. He published his work in a book titled *Electric Waves: Being Researches on the Propagation of Electric Action With Finite Velocity Through Space.* These experiments greatly expanded the field of electromagnetic transmission, and other scientists in the field eventually further developed the Hertz antenna receiver.

Hertz also found that radio waves could be transmitted through different types of materials and were reflected by others. This discovery ultimately led to the invention of radar. He even paved the way for wireless communication, although he never recognized that important aspect of his experiments.

In recognition of his contributions, the hertz designation has been an official part of the international metric system since 1933. It is the term used for units of radio and electrical frequencies.

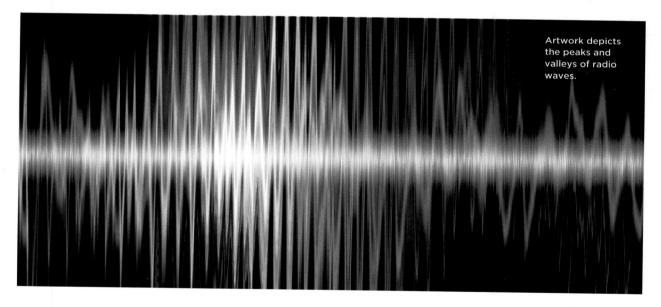

Artwork depicts the peaks and valleys of radio waves.

>> **FACT:** Radio waves have helped develop such wireless devices as telephones and computer keyboards.

# 19 Large-Scale Electric Supply Network (1880)

Inventor Thomas Edison was the first person to devise and implement electric power generation and distribution to homes, businesses, and factories, a key milestone in the development of the modern industrialized world. Edison patented this system in 1880 in order to capitalize on his invention of the electric lamp—he was nothing if not a shrewd businessman. On December 17, 1880, he founded the Edison Illuminating Company, headquartered at Pearl Street Station in New York City. On September 4, 1882, Edison switched on his Pearl Street generating station's electrical power distribution system, which provided 110 volts of direct current to about 60 customers in Lower Manhattan.

Although Edison lost the so-called War of the Currents that followed—with the consequence that alternating current became the system through which electrical power was distributed—his power-distribution system is still significant for a few reasons. It established the commercial value of such a system, and it helped to stimulate advances in the field of electrical engineering, as people began to see the field as a valuable occupation. For example, American electrical engineer Charles Proteus Steinmetz, through his work in alternating current, made possible the expansion of the electric power industry in the United States by formulating mathematical theories for engineers who were designing electric motors for use in industry.

In 1880 Thomas Edison produced a long-lasting source of light by using a small filament and a vacuum inside a glass globe.

>> **FACT:** The electric power system in the United States is the largest in the world.

# 20 Boolean Logic (1854)

**C**omputer operations are based on one thing: determining whether a gate, or switch, is open or closed, which is often signaled by the numbers 0 and 1. This is the essence of the Boolean logic that underlies all of modern computing. The concept was named after English mathematician George Boole, who defined an algebraic system of logic in 1854. Boole's motivation to create this theory was his belief that the symbols of mathematical operation could be separated from those of quantity and could be treated as distinct objects of calculation.

Nearly a century later, Claude Shannon showed that electric circuits with relays were perfect models for Boolean logic, a fact that led to the development of

the electronic computer. Computers use Boolean logic to decide if a statement is true or false (this has to do with a value, not veracity). There are three basic gates: AND, OR, and NOT. The AND operation says that if and only if all inputs are on, the output will be on. The OR operation says that if any input is on, the output will be on. The NOT operation says that the output will have a state opposite to that of the input.

>> **FACT:** Though invented in the 19th century, Boolean logic forms the basis of most 21st-century Internet searches.

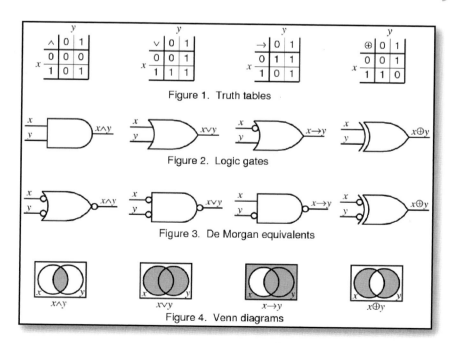

Figure 1. Truth tables

Figure 2. Logic gates

Figure 3. De Morgan equivalents

Figure 4. Venn diagrams

0=OPEN
1=CLOSED

The algebraic tradition in logic, developed by English mathematician George Boole (above), today finds application in various kinds of decision-making (left).

Origami-like designs drive projects at the micro and even nano scale.

Foreleg      Neck and head

Tail

Rear leg

Rear leg      Foreleg

# 21 Programmable Matter

Programmable matter, which was first introduced by Massachusetts Institute of Technology (MIT) researchers Tommaso Toffoli and Norman Margolus in a 1991 paper, is quickly becoming a reality, albeit on a relatively small scale.

Programmable matter is matter that can simulate or form different objects as a result of either user input or its own computations. Some programmable matter is designed to create different shapes, while other matter, such as synthetic biological cells, is programmed to work as a genetic toggle switch that signals other cells to change properties like color or shape. The potential benefits and applications of programmable matter—in particular, the possibility of using it to perform information processing and other forms of computing—have created a lot of excitement in the research world.

Since the publication of Toffoli and Margolus's paper, much work has been done to fulfill the potential they predicted. In 2008, for example, the Intel Corporation announced that its researchers had used programmable matter to develop early prototypes of a cell phone device at the centimeter and millimeter scale. In 2009 the Defense Advanced Research Projects Agency of the U.S. Department of Defense reported that five different teams of researchers from Harvard University, MIT, and Cornell University were making progress in programmable matter research. In 2010 one of these teams, led by Daniela Rus of MIT, announced its success in creating self-folding sheets of origami. The Defense Advanced Research Projects Agency harbors hopes for futuristic applications of the concept. Its members envision, for example, a soldier equipped with a lightweight bucket of programmable material that could be fashioned, on the spot, into just about anything he or she needed.

• **1** NANOLITHOGRAPHY has been used to create nano-size circuits, which could be used in programmable matter.

• **12** In 1959 physicist Richard Feynman predicted the possibility of **NANOTECHNOLOGY**, which will play a crucial role in the creation of programmable matter.

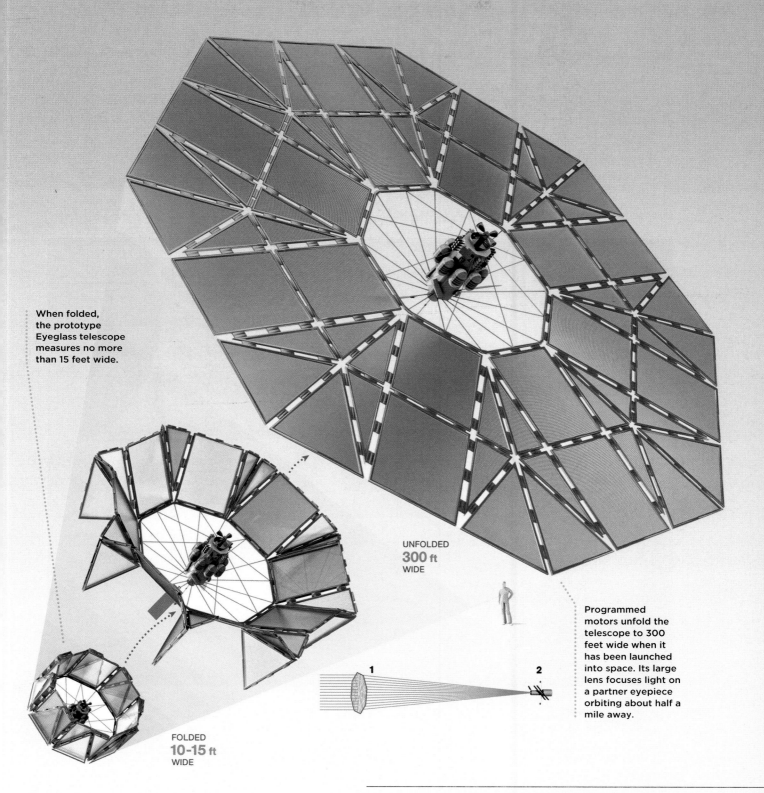

When folded, the prototype Eyeglass telescope measures no more than 15 feet wide.

UNFOLDED
**300 ft**
WIDE

FOLDED
**10-15 ft**
WIDE

Programmed motors unfold the telescope to 300 feet wide when it has been launched into space. Its large lens focuses light on a partner eyepiece orbiting about half a mile away.

# 22 Difference Engine (1822)

Charles Babbage, an English mathematician, philosopher, inventor, and mechanical engineer, realized in 1822 that an engine of sorts could be programmed with the use of paper cards that stored information in columns containing patterns of punched holes. Babbage saw that a person could program a set of instructions by way of punch cards, and the machine could automatically carry out those instructions. The intended use of Babbage's difference engine was to calculate various mathematical functions, such as logarithms. Although it was never completed, it is considered one of the first general-purpose digital computers.

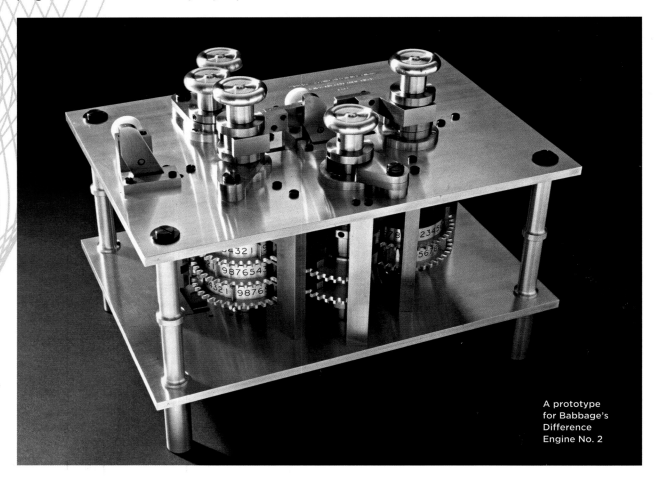

A prototype for Babbage's Difference Engine No. 2

>>**FACT:** In 2011 British researchers began constructing the engine Babbage designed but never built.

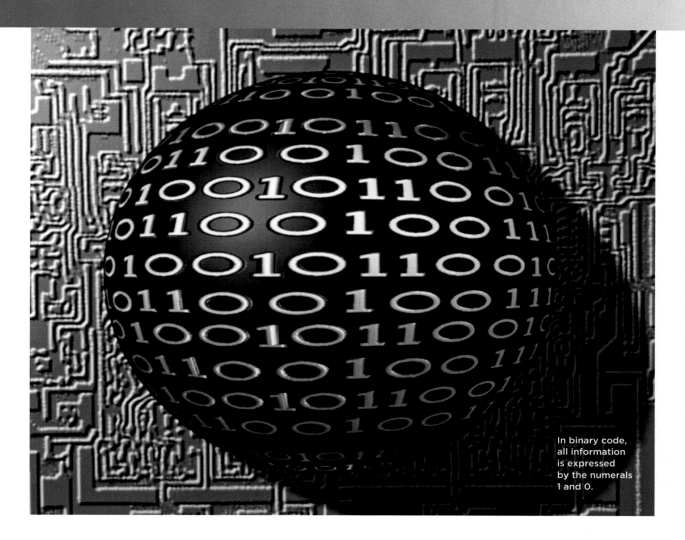

In binary code, all information is expressed by the numerals 1 and 0.

# 23 Binary Numbers/ Binary Code (1697)

I n the 1670s German philosopher and mathematician Gottfried Wilhelm Leibniz—whose accomplishments include the invention of calculus and many important advances in math, logic, and science—invented an arithmetic machine that could multiply as well as add. Leibniz also saw how his machine could be altered to use a binary system of calculation—a concept that is at the heart of digital computing.

In Leibniz's system, the term *binary* refers to a number system whereby all values are expressed with the numbers 1 and 0. The binary system can be best understood by contrasting it with today's base 10 system, which expresses numbers using 0 through 9. In base 10, the number 367, for example, represents $3 \times 100 + 6 \times 10 + 7 \times 1$. Each position in the numeral 367 represents a power of ten, beginning with zero and increasing from right to left. In the binary, or base 2, system, each position represents a power of two. So, in binary, 1101 represents $1 \times 2^3 + 1 \times 2^2 + 0 \times 2^1 + 1 \times 2^0$, which is equal to $8 + 4 + 0 + 1$, or 13.

# 24 Modern Number System (800)

The Hindu-Arabic numeral system upon which the modern number system is based was probably developed in the ninth century by Indian mathematicians, adopted by Persian mathematician Al-Khwarizmi and Arabic mathematician Al-Kindi, and spread to the Western world by the high Middle Ages.

The characteristics of the modern number system are the concepts of place values and decimal values. The place value system indicates that the value of each digit in a multidigit number depends on its position. Take the number 279, for example. According to the place value system, the 2 represents hundreds, the 7 represents tens, and the 9 represents ones. Thus the number appears as 279. Meanwhile, the related decimal system presents numbers in increments of ten. In other words, each place value is ten times the value of the place before it. The decimal system allows mathematicians to perform arithmetic with high numbers that would otherwise be extremely cumbersome to manipulate.

Computers make use of the positional numbering system. Since a computer uses a small amount of memory to store a number, some numbers are too large or too small to be represented. That is where floating-point numbers come in. The decimal point can "float" relative to

the significant digits in a number. For example, a fixed-point representation that has seven decimal digits with two decimal places can represent the numbers 12345.67, 123.45, 1.23, and so on, whereas the same floating-point representation could also represent 1.234567, 123456.7, 0.00001234567, and so on.

A telephone dial displays an age-old way of writing numbers.

A wooden abacus, considered the first personal computer

# 25 Abacus (3000 B.C.)

The invention of the abacus is credited to the Chinese about 5,000 years ago. Its use quickly spread throughout the world. Whether counting flocks of sheep, decanters of wine, or sacks of grain, the abacus gave people the ability to keep track of numbers as they did their figuring. The abacus makes use of the place value system, in that each row of stones, beads, or marbles stands for a unit number—ones, fives, tens, and so on.

In a sense, the abacus can be considered one of the first computers. While today's computers do the actual calculating, a proficient abacus user can perform simple calculations almost as quickly.

>>**FACT:** The abacus is the foundation of all computing devices, from the Leibniz calculator to the PC.

# 02

# ENGINEER THE BODY

Afflictions of the body—sickness, injury, and disability—are a universal human concern transcending boundaries of space and time. Against this dark backdrop, the history of medical science is a narrative of progress and hope.

Driven by a spirit of inquiry, medicine has radically transformed experience. For much of the human past, for example, the human body was a terra incognita, as much the object of fantasy as of heaven or hell. Imagine the fascination with which early modern Europeans pored over the first accurate, detailed, full-color renderings of the stomach and bowels, or of the nervous system's delicate filigree.

English polymath Robert Hooke first looked through a microscope and observed the strange structures he called cells in the 1600s. This drive to explore the intricacies of biology led to the discovery of increasingly minute and

# ING

finally invisible shapes—bacteria, viruses—whose discovery in turn reshaped the human life cycle.

By the end of the 20th century researchers also had uncovered the structure and function of DNA and had completed a map of the human genome. Today, scientists are creating genetically modified organisms and exploring the field of epigenetics, which aims to understand hereditary changes in gene expression without changes to the DNA itself. Innovators in the field of synthetic biology are harnessing knowledge of microbiology, biochemistry, and other fields to create completely new forms of life.

If the challenges scientists face today—the quest to engineer replacement organs, or to enlist stem cells to repair damaged hearts and brains—seem all but insurmountable, a backward glance at medicine's past accomplishments inspires confidence in its future.

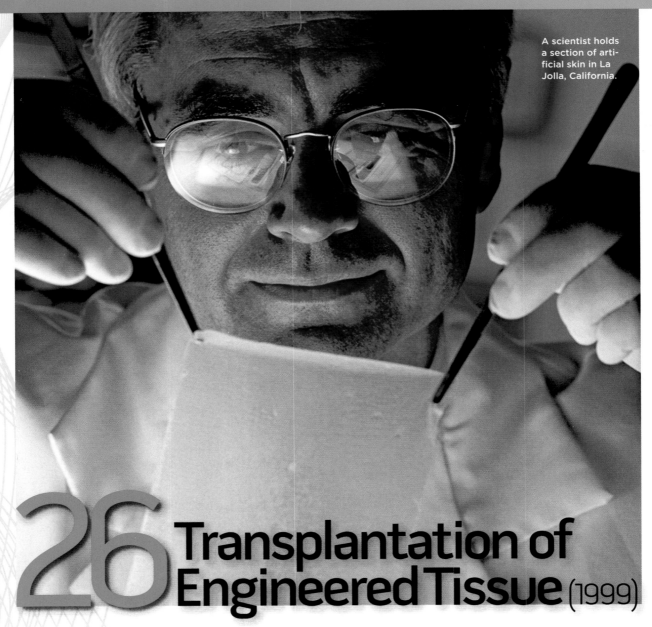

A scientist holds a section of artificial skin in La Jolla, California.

# 26 Transplantation of Engineered Tissue (1999)

Against the backdrop of years of basic science, the process of building the first functioning lab-grown human organ could seem deceptively simple. First, researcher Anthony Atala and colleagues at the Wake Forest Institute for Regenerative Medicine collected muscle and bladder cells from their patient and cultured them in the lab. The scientists formed a bladder-shaped scaffold of biodegradable material, covered this mold with the patient's own muscle cells, and lined it with bladder cells. Then they grafted their construction onto the patient's own dysfunctional bladder. More than six years later, in 2006, the group announced a triumph to the world: The first bladder they had implanted, along with a handful of others, was working.

>> **FACT:** A beagle was the first recipient of an artificial bladder grown from the animal's own bladder cells.

# 27 Gene Therapy (1993)

Devastating inherited diseases have played a key role in the nascent science of gene therapy by dealing researchers early triumphs—and painful setbacks.

In 1990 scientists at Maryland's National Institutes of Health launched gene therapy's first human trial when they treated two girls with severe combined immunodeficiency, or SCID. They collected the children's defective white blood cells, inserted normal genes into them, and then returned the cells to the patients. Though they required ongoing treatment, the girls developed normal immunity.

In 1999, though, the public learned of a gene-therapy failure in the death of 18-year-old Jesse Gelsinger, whose inherited liver disease was being treated in a University of Pennsylvania clinical trial. A government investigation found a number of ethical lapses in the trial's conduct, including a failure to inform subjects fully about risks.

Then, in 2000, a team from Paris reported results of a gene-therapy trial in ten boys with a different form of SCID. The boys had received transplants of their own "corrected" bone-marrow cells. Tragically, two years later, the team reported that one boy, then another, had developed a leukemia-like condition, apparently because the virus used to deliver the boys' corrected DNA had lodged near a gene that regulates cell growth. The alarming news led to the halting of not only the French trial but also studies in Germany, Italy, and the United States.

More recently, though, small gene-therapy experiments have produced positive results. Reported in 2010, a University of Pennsylvania gene-therapy study showed improvement in seven of eight subjects with HIV.

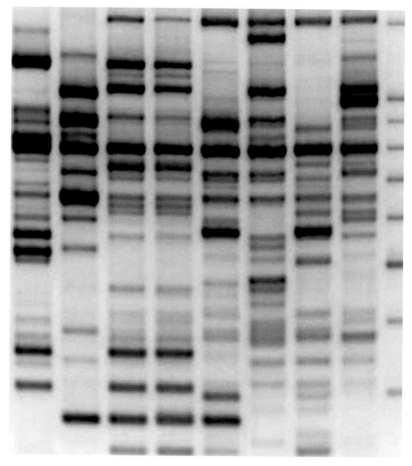

A DNA gel sequence (above) is used primarily for analysis. A conceptual image shows a pill capsule (left) that might be used for gene therapy.

# 28 Genome Sequencing (1977)

Genome sequencing is a process that "reads" the order of DNA nucleotides in a genome—that is, the order of the base pairs A–C and G–T, which make up an organism's DNA. The 1977 genome-sequencing work of Frederick Sanger, one of the most famous scientists in the history of genetics, led to the basic research that made the Human Genome Project possible. (Sanger is also known for developing methods of amino acid sequencing that enabled him to determine the complete amino acid sequence of insulin before anyone else in 1951.)

The system used to determine the sequence of DNA, called Sanger sequencing, involves separating fluorescent-labeled DNA fragments based on the length of a polyacrylamide gel. The base at the end of each fragment is identified by how it reacts to a specific kind of dye.

Sanger used this technique to sequence the DNA of bacteriophage fX174, a viral genome with 5,368 base pairs. He discovered that there was overlap among the genes in some areas with respect to coding, a finding that enabled geneticists to analyze longer strands of DNA more rapidly and with greater accuracy than before. In 1980, for this achievement, Sanger was awarded a second Nobel Prize in chemistry, which he shared with Walter Gilbert and Paul Berg. Sanger had received his first Nobel Prize in 1958 for work on the structure of proteins.

>> **FACT**: The genomes of more than 180 organisms have been sequenced since 1995.

Harvard scientist Walter Gilbert studies a DNA sequencing autoradiogram.

The genetic material of a tomato and a kiwi has been modified to produce a hybrid fruit.

# 29 Genetic Engineering (1973)

Genetic engineering, also called genetic modification, involves directly manipulating an organism's genetic material. It uses recombinant DNA, in which two or more genetic sequences are combined in a way that would not ordinarily occur in nature. The origin of this field is attributed to American biochemists Herbert Boyer and Stanley Cohen, who invented the technique of DNA cloning. The first genetically engineered organisms were bacteria in 1973, followed by mice in 1974. Since then, scientists have used genetic engineering techniques in research, biotechnology, medicine, and other fields.

In this process, desired genetic material is isolated and copied with care to ensure that the genes will express themselves correctly. The new genetic material is placed into a host genome. This is the most common form of genetic engineering. Other forms include techniques that are used to target and remove specific genes.

# 30 Regenerative Medicine

The burgeoning field of regenerative medicine seeks nothing less than to provide patients with replacement body parts. Here, the parts are not steel pins and such. They are the real thing: living cells, tissue, and even organs.

Regenerative medicine is still a mostly experimental enterprise, with clinical applications limited to such procedures as growing sheets of skin to graft onto burns and wounds. But the prospects go much further. As long ago as 1999, a research group at North Carolina's Wake Forest Institute for Regenerative Medicine implanted a patient with a laboratory-grown bladder. The team has continued to generate an array of other tissues and organs, from kidneys to salivary glands to ears.

In 2007 a team led by orthopedic surgeon Cato Laurencin, then at the University of Virginia, reported on a tissue-engineered ligament that could allow patients to recover more quickly and fully from one of the most common types of knee injury—the torn anterior cruciate ligament (ACL). Laurencin's ACL was made of braided synthetic microfibers "seeded" with actual ACL cells. Tested in rabbits, the scaffold, a supporting framework, promoted new blood vessel and collagen growth within 12 weeks.

Also working in animal models, other researchers have made important strides in testing therapies based on stem cells, which multiply rapidly and can differentiate into a variety of cell types. These repair cells may eventually be deployed to regrow cardiac muscle damaged by heart attack, or to replace nerve cells in victims of spinal-cord injury.

The genesis of this approach reaches back to the early 20th century and the first successful transplantations of donated human soft tissue, bone, and corneas. Much as transplant medicine has progressed, it suffers from an intractable problem that regenerative medicine might one day sweep aside: There are not enough donor organs for people who need them, so many patients die while waiting for an organ. Another advantage of regenerative medicine is that the body's immune system will not reject tissues grown from a patient's own cells.

At North Carolina's Wake Forest Institute for Regenerative Medicine, cells are grown to match organs or body parts, such as ears.

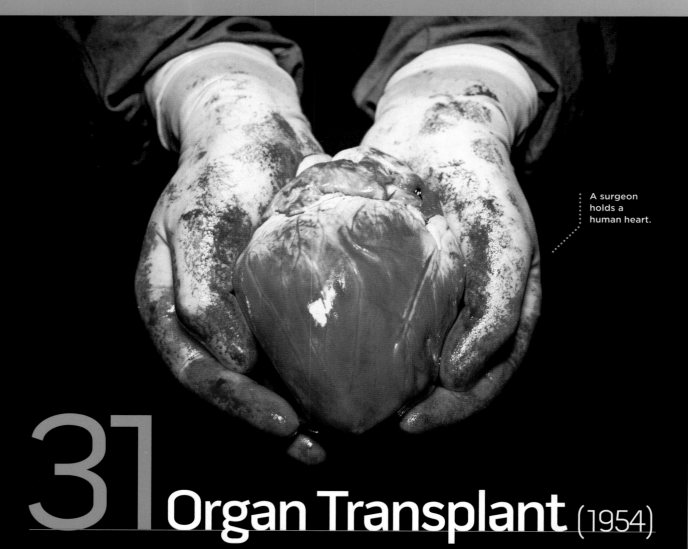

A surgeon holds a human heart.

# 31 Organ Transplant (1954)

At Boston's Peter Bent Brigham Hospital in 1954, doctors led by surgeon Joseph E. Murray performed the world's first successful organ transplant, removing a healthy kidney from one man and sewing it into his identical twin, a victim of severe renal disease.

The surgical challenges were considerable, and the procedure raised a new ethical question: For the first time one patient (the donor) would submit to a highly invasive procedure solely to benefit another patient.

But the surgeons did not so much overcome as avoid the most substantial obstacle to success in that first transplant. The team understood from earlier work on skin grafts that the closer the relation between donor and recipient, the better the chances a graft would survive. Transplantation between genetically identical twins was a chance to demonstrate that if doctors could overcome immune rejection, they could save lives with organ transplantation.

The surgery's success encouraged a spate of attempts around the world to transplant organs and, perhaps even more important, a period of intensive experimentation in immunology. Researchers tried a number of approaches, including bombarding the recipient's immune system with radiation and suppressing it with experimental drugs. Murray's own lab worked with Imuran, the first drug approved in 1963 for immunosuppression in transplant surgery. The first two patients to receive Imuran under Murray's care died from its toxicity, but the third received the first successful organ transplant from an unrelated donor.

# 32 Structure of DNA (1953)

Perhaps there is no symbol of science more iconic than the DNA double helix. The 1953 discovery of this structure by scientists James Watson and Francis Crick at Cambridge University was nothing short of monumental. DNA, or deoxyribonucleic acid, is the substance that contains the genetic instructions for all living things, whether human, horse, housefly, or bacterium. Prior to Watson and Crick, the existence of DNA was known, although some scientists were skeptical that it contained genetic material. What remained to be determined was what DNA actually looked like.

Watson and Crick created models out of sticks and balls similar to Tinkertoys. They originally operated

under the erroneous assumption that DNA was a triple helix. Although Watson and Crick are credited with the discovery of the structure of DNA, they built on the work of others—particularly British biophysicist Rosalind Franklin, who in 1952 used a painstaking technique called x-ray diffraction to make a famous three-dimensional image of DNA. Franklin was among the first to speculate on DNA's physical structure.

DNA's twisting helix (below) was discovered by James Watson (above, at left) and Francis Crick (above, at right).

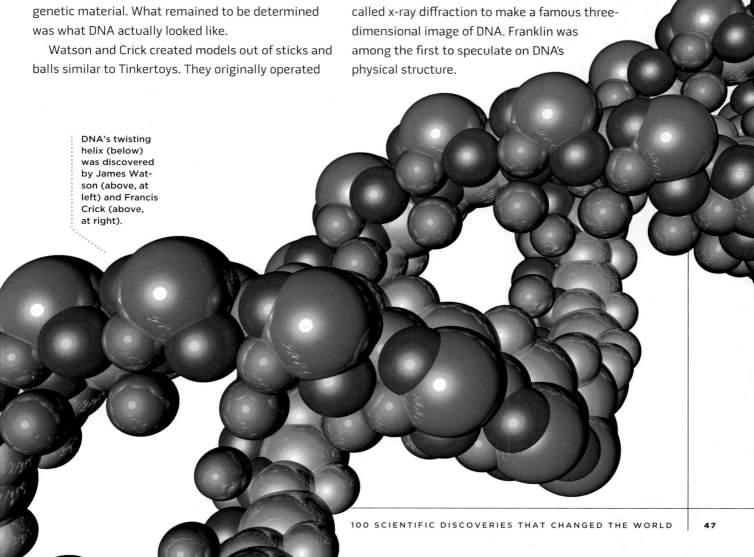

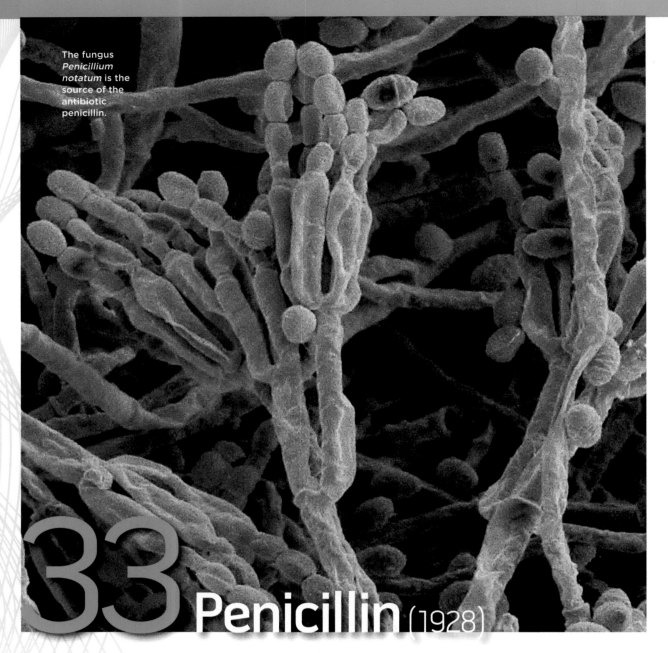

The fungus *Penicillium notatum* is the source of the antibiotic penicillin.

# 33 Penicillin (1928)

When Scottish biologist Alexander Fleming returned from his 1928 summer vacation, he noticed a mold growing in one of his dishes of cultured bacteria. Surrounding the mold was a bacteria-free nimbus, as if the mold's juice were toxic to the pathogenic *Staphylococcus* bug.

Fleming understood the potential of the agent he named penicillin. He tested it against an array of disease-causing organisms, with promising results. But the mold was hard to grow, and its active ingredient was hard to isolate—difficulties that would be overcome a decade later when, in the context of World War II, Allied governments mounted an intensive effort to produce a potent weapon against infection.

>> **FACT**: Penicillin works by preventing the development of bacterial cell walls.

# 34 Chromosomes Carry Genes (1910)

Perhaps no early scientist spent more time peering at fruit flies through microscopes than American biologist Thomas Hunt Morgan. In 1910 Morgan noticed a single red-eyed fly among his white-eyed specimens. He conducted a series of breeding experiments to study how this mutation would be inherited. He found that the white-eyed trait was recessive—a single red-eyed parent invariably meant red-eyed offspring. He also speculated that it was located on the X chromosome, since only males (with no second X chromosome to counteract the white-eyed form of the gene) displayed the characteristic.

Morgan had discovered that specific genes are carried on specific chromosomes. He was building on work pioneered by American geneticist Walter Sutton and German biologist Theodor Boveri, who in 1902 independently came to the conclusion that chromosomes carry genetic material. These were important milestones in the understanding of inheritance and genetics.

>> **FACT:** Morgan was awarded the 1933 Nobel Prize in medicine for his work with chromosomes and heredity.

Experiments with red-eyed fruit flies (above right) revealed that specific genes are carried on specific chromosomes (right).

# 35 Mitosis (1879)

The German cytologist Walther Flemming is credited with the discovery of mitosis in 1879. Flemming had developed a new staining technique that allowed him to identify and observe chromosomes in greater detail than before. As a result he was able to observe cell division, considered to be one of the great scientific discoveries of all time.

Except for sex cells, all cells—those responsible for growth, development, and cellular repair, for example—divide by mitosis. Sex cells—eggs and sperm—divide by meiosis, which was explained by German biologist August Weismann in 1890.

>> **FACT:** The word "mitosis" comes from the Greek word *mitos,* meaning "thread."

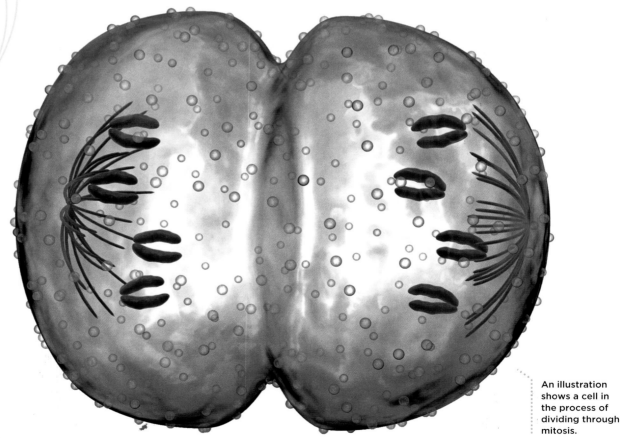

An illustration shows a cell in the process of dividing through mitosis.

Research with pea plants led Gregor Mendel to outline the rules of trait inheritance.

# 36 Mendelian Genetics (1865)

Although the study of genetics began before Gregor Mendel came on the scene, it is Mendel's innovative work with pea plants in the 1860s, published in his 1865 seminal paper "Experiments in Plant Hybridization," that furthered the understanding of inheritance. Mendel's work led to three principles that can be called rules for trait inheritance:

(1) An organism's phenotype (what it looks like) cannot be used as a basis for determining its genotype (genetic structure).

(2) The law of segregation states that genes retain their individuality and do not combine to form a new, blended trait. For example, breeding a black-and-white dog with a brown-and-white dog will produce some dogs that are black and white, and some that are brown and white, but no dogs that are brown, black, and white.

(3) The law of independent assortment states that every gene has both dominant and recessive forms, which is why traits can skip generations.

# 37 Genetically Modified Organisms

Although genetically modified organisms have been around for a while (the first experiments were conducted in 1978), scientists are just now gaining a better understanding of how these organisms may affect people's lives. A genetically engineered organism is created using recombinant DNA technology, in which DNA molecules from different sources are combined into one molecule to create a new set of genes. This modified DNA is inserted into an organism, giving it genes that nature did not.

Genetically modified organisms are used in biological and medical research, production of pharmaceutical drugs, gene therapy, and agriculture. The general public is perhaps most familiar with genetically modified organisms in the form of crops such as corn that are altered to be "naturally" resistant to pests and herbicides while producing the highest yield possible. However, bacteria were the first organisms to be genetically modified due to their simple structure. These bacteria are now used for several purposes: to produce insulin protein to treat diabetes, as clotting factors to treat hemophilia, and as human growth hormones to treat various forms of dwarfism, to cite three examples.

The use of genetically modified organisms has sparked significant controversy. Some people view the technology as unacceptable meddling with biological states and processes that have evolved naturally over long periods, while others are concerned about the limitations of modern science to grasp fully all of the potential negative ramifications of genetic manipulation. To date, no studies have shown a documented link between adverse health effects and the consumption of genetically modified foods, yet environmental groups in many countries, especially those of the European Union, still discourage consumption of genetically modified foods by claiming these foods are unnatural and unsafe. Clearly more research is needed before it is known for certain whether the benefits of this technology outweigh the risks.

• 28 With GENOME SEQUENCING technologies, scientists can glimpse the entire genomes of the organisms they wish to modify.
• 29 GENETIC ENGINEERING forms the foundation for the genetic modification of organisms.
• 32 GENETICALLY MODIFIED ORGANISMS wouldn't be possible without the discovery of the structure of DNA.
• 34 The discovery of CHROMOSOMES AS GENE CARRIERS was a crucial early step in the science of genetics.

BREAKT

Researchers imported the gene for fluorescence from a jellyfish into a virus and then introduced the virus into a mouse egg. The result: a mouse that glows in the dark.

Glowing mice may interest cats, but this experiment has more profound implications. Using the same technique, doctors may soon tag cancer cells in a patient's body and then track their growth and movement as never before.

# HROUGH

# 38 Theory of Evolution (1859)

*The Origin of Species* outlines Darwin's groundbreaking theories.

Perhaps no other scientific theory has sparked such debate and controversy as Charles Darwin's theory of evolution. Prior to Darwin, the long-standing beliefs were that Earth was no more than about 6,000 years old and that species had no relationship to one another. Humans were considered distinct and superior to all other organisms.

In 1831 the British naturalist embarked on his now famous voyage on the H.M.S. *Beagle,* which took him around the world and allowed him to study a wide variety of flora and fauna. Darwin returned to England in 1836 and began compiling his observations into his book *On the Origin of Species,* which was published in 1859. In this book Darwin presents his theory of natural selection, whereby organisms that are able to adapt successfully to their environments have more offspring than those who are less successful at adaptation. The traits of those better-adapted individuals increase in the population, and over time the species evolves in response to environmental factors.

Through studying orchids, finches, and tortoises, Darwin came to the sudden realization that "it is absurd to talk of one animal being higher than another." This flew in the face of religious doctrine, whereby humankind was created in the image of God. And so the debate began.

Darwin and natural selection may be inextricably linked in the history books, but he was not the only one to propose a theory of evolution. British naturalist Alfred Russel Wallace, a colleague of Darwin's, presented similar findings in a paper titled "On the Tendency of Varieties to Depart Indefinitely from the Original Type."

Modern elephants (top and center) are descended from extinct species (bottom).

# 39 Germ Theory of Disease (1859)

Puzzling out the true nature of infectious disease was perhaps the most important scientific project of the 19th century. Observation over many generations had led to various erroneous but persistent theories. At mid-century the prevailing belief was that microbes could simply spring to life out of inanimate matter (spontaneous generation), and that disease, if transmitted at all, was not spread from host to human host on droplets of sputum and the like, but was carried on a miasma of noxious, foul-smelling air.

Scientists overturned these ideas in fits and starts. Dutch amateur lens grinder Anton van Leeuwenhoek first spied various bacteria in the 1670s. It was not until the 19th century that Hungarian physician Ignaz Semmelweis developed, tested, and proved his theory that a much-dreaded childbed fever was being transmitted on the hands of physicians, who, fresh from their work at the autopsy table, attended women giving birth in Semmelweis's Vienna hospital in the 1840s. Unfortunately, his urging that doctors clean their hands before treating patients was met with indifference and outright hostility.

By the time Louis Pasteur conducted his famous experiment in 1859, the idea that germs could infect and sicken a person was familiar, but it was locked in a pitched battle with competing theories. Pasteur's work all but decided the contest. His flasks of broth covered with filters or fitted with narrow, downward-curving stems admitted air but excluded the tiniest particles—and remained free of bacterial growth. The enemy, though invisible to the naked eye, lay in medicine's sights.

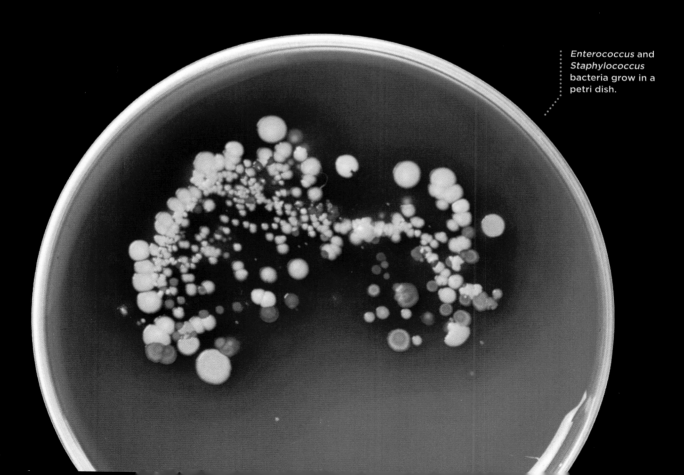

*Enterococcus* and *Staphylococcus* bacteria grow in a petri dish.

# 40 Anesthesia (1846)

It is difficult to imagine how anyone endured a tooth extraction—much less a limb amputation—before the age of anesthesia. A key qualification for surgeons back then was speed; in addition, attendants had to be strong enough to restrain the patient's agonized writhing. By the early 1840s physicians had no better methods for dulling surgical pain than to stupefy the patient with a kind of hypnosis known as mesmerism—or, indeed, with whiskey.

A turning point came in 1844 when Connecticut dentist Horace Wells attended a demonstration of nitrous oxide's intoxicating effects. Wells thought of using the so-called laughing gas during invasive procedures when he noticed one participant had scraped his leg badly but had suffered no pain. Like others before him—the anesthetic properties of nitrous oxide had been recognized for decades—Wells failed to develop the idea beyond an unsuccessful demonstration at Massachusetts General Hospital (the patient cried out as his tooth was pulled).

The eureka moment would be left to Wells's associate, dentist William Morton. In 1846, while working with ether, Morton seized the opportunity to work on a patient with an infected tooth. He etherized the man, extracted his bicuspid, and, when the patient awakened a few minutes later, announced to the patient's astonishment that the procedure was done. Venturing another demonstration at Massachusetts General, Morton anesthetized a patient undergoing surgery on a neck tumor. After the procedure the initially skeptical surgeon turned to the audience and announced, "Gentlemen, this is no humbug."

>> **FACT**: Queen Victoria of England used chloroform to deliver her eighth and ninth children in 1853 and 1857.

An anesthetic bottle and case from the early 20th century

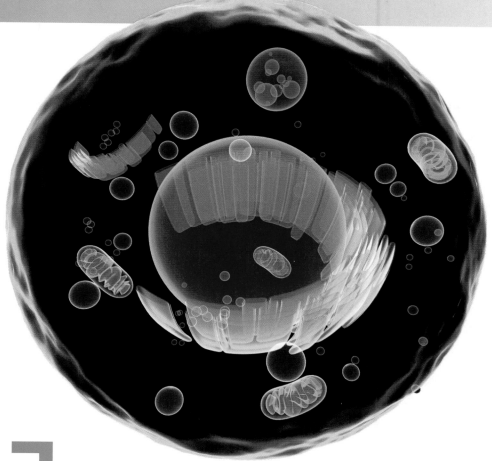

Research in the 19th century revealed that the cell is the smallest form of life.

# 41 Cell Theory (1838)

The observations of several scientists in the early 19th century led to what is called cell theory—the idea that cells are the building blocks of all life. When Germans Matthias Schleiden and Theodor Schwann presented their findings in 1838, the study of biology changed forever. According to cell theory, cells are the smallest forms of life and all living things are composed of them. Furthermore, only preexisting cells can create cells; new cells do not arise spontaneously or come from another source.

Schleiden based his research on the work of Scottish botanist Robert Brown, who had discovered the cell nucleus. It was Schleiden, however, who understood the true importance of the nucleus and its role in the development of the complete cell. At the same time, Schwann was studying animal cells and trying to work out a puzzle: Why did certain structures in animal and plant cells look so similar?

It was Schleiden's observations of the nucleus that gave the biologists the answer. Because both plant and animal cells contain this structure, Schleiden proposed that cells must be the building blocks of life. The primary difference lies in the fact that plant cells have rigid walls and animal cells do not. This is significant because it gives animal cells the malleability to take on various shapes. Meanwhile, the chloroplasts in plant cells enable plants to use sunlight for food in a process called photosynthesis. Animal cells do not have this ability; they need to absorb nutrients from other cells in a process called phagocytosis.

# 42 Smallpox Vaccine (1796)

It was during his apprenticeship that English country physician Edward Jenner overheard a local girl pass on a folk belief. She would never suffer the ugly scars of smallpox, she exulted, because she had already had cowpox, a minor affliction common to dairymaids.

Years later, in 1796, Jenner collected material from a cowpox lesion on the arm of a local milker and rubbed it into a small scoring in the skin of an eight-year-old boy. Eight weeks later Jenner exposed the boy to smallpox. The child remained well. Cowpox was similar enough to smallpox to induce immunity. Over the next several years, Jenner published further experiments, supplying cowpox material and promoting his techniques to physicians around the world. Vaccination with cowpox became mandatory in Bavaria, Denmark, Prussia, and finally Britain, in 1853. American states also began to require the vaccine at mid-century.

Jenner's innovation was a critical step toward reducing the threat of infectious disease. It made early childhood, in particular, vastly safer.

In the 1960s, worldwide health organizations launched a program to eradicate smallpox from the globe. The last natural case occurred in Somalia in 1977 and the World Health Assembly certified the disease's eradication in 1980. However, millions of doses of the vaccine are still available should the virus ever reappear.

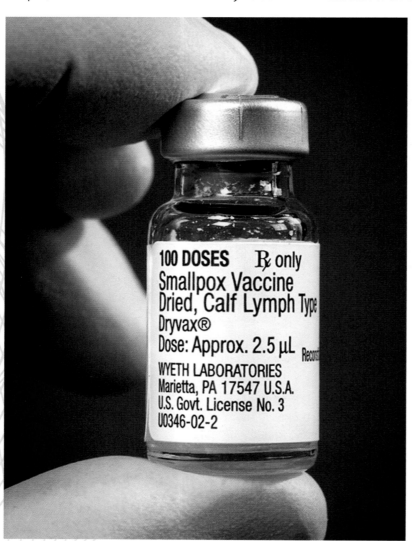

100 DOSES ℞ only
Smallpox Vaccine
Dried, Calf Lymph Type
Dryvax®
Dose: Approx. 2.5 μL
WYETH LABORATORIES
Marietta, PA 17547 U.S.A.
U.S. Govt. License No. 3
U0346-02-2

Worldwide smallpox vaccinations (vaccine, above) have wiped out the deadly virus (above right).

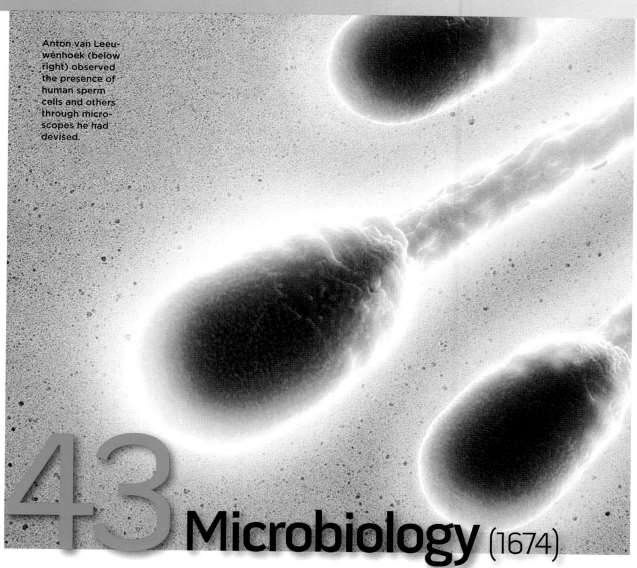

Anton van Leeuwenhoek (below right) observed the presence of human sperm cells and others through microscopes he had devised.

# 43 Microbiology (1674)

Known as the father of microscopy because of his advances in the design of microscopes, Anton van Leeuwenhoek is noted for his observations of protozoa and bacteria. The Dutch naturalist was the first to observe and document these organisms in 1674, and his findings paved the way for the field of microbiology and opened a whole new world to the scientists of his day. In addition to protozoa and bacteria, Leeuwenhoek observed yeast cells, blood cells, sperm cells, and more.

A man possessing an innate, insatiable curiosity, Leeuwenhoek wrote in 1716, "[M]y work . . . was not pursued in order to gain the praise I now enjoy, but chiefly from a craving after knowledge . . . [W]henever I found out anything remarkable, I have thought it my duty to put down my discovery on paper, so that all ingenious people might be informed thereof."

>> **FACT:** Anton van Leeuwenhoek was named a Fellow of the Royal Society of London for his discoveries.

# 44 Nanomedicine

The prefix "nano-" means "one-billionth." Thus, a nanometer is a billionth of a meter, which measures about half the diameter of DNA's famous double helix. This is the scale at which the field of nanotechnology aspires to manipulate matter. This manipulation is made possible—or at least imaginable—by the emergence in the 1980s of high-powered microscopes that permitted scientists to see such tiny particles for the first time.

The ultimate medical application of nanotechnology would be to engineer minuscule machines capable of repairing cells. In the meantime, scientists are working to develop novel drug-delivery systems in which active ingredients ride straight to their target cells on nanoparticles. Scientists are also working to develop diagnostic tools featuring nanoparticles that bind to or otherwise mark disease-indicating molecules. Other applications use the tiny particles to attack viruses or inflammation-causing free radicals preferentially.

One nanotech drug candidate now being tested in humans, Maryland-based CytImmune's Aurimune, is designed to overcome a notorious problem in cancer treatment: Drugs that kill cancer cells often are toxic to the entire body. Aurimune is a 27-nanometer particle of gold carrying a tumor-killing molecule as well as one designed to hide the particle from the immune system. The idea—and early studies suggest that it just might work—is that the drug flows freely through the bloodstream; small enough to escape through the leaky blood vessels in and around tumors, it deposits its cargo there and spares healthy tissue.

In another intriguing example, researchers at 11 sites across Europe and Australia are working on a wound dressing coated with nanocapsules, which contain both an antibiotic and a dye. These exquisitely tiny capsules are designed to burst when exposed to a toxin secreted by pathogenic bacteria; the dressing immediately begins to treat the wound infection while changing color to alert doctors. The project is scheduled to wind up in 2014, by which time scientists hope to have a prototype ready for industrial production.

For all its promise, nanomedicine also raises new controversies. Perhaps chief among these is a concern that particles so tiny that they readily enter cells and cross the blood-brain barrier may carry long-term risks that are not well understood.

- **2 CARBON NANOTUBES** will form the foundation of some of the technologies of nanomedicine.
- **4** The properties of nanoparticles called **FULLERENES** make them great candidates for use in nanomedical technologies.

- **49** & **50** The ideas of **EXPERIMENTAL MEDICINE** and the **SCIENTIFIC STUDY OF MEDICINE** are fundamental to the new field of nanomedicine.

The stylus of an atomic force microscope moves an atom and creates a new nanometric arrangement.

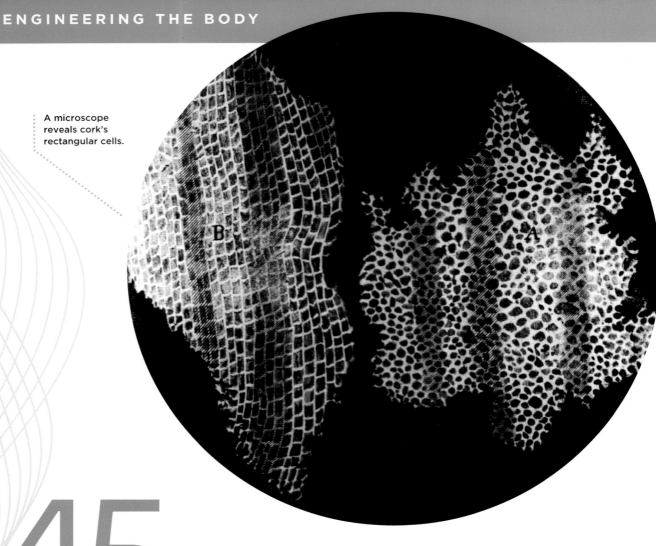

A microscope reveals cork's rectangular cells.

# 45 Cells (1665)

In 1665, while looking at thin slivers of cork through a microscope, Robert Hooke noticed small holes, or what he coined cells. The Englishman later said that the cavities reminded him of monks' quarters (hence the name). He believed these cells had once been containers for "noble juices" or "fibrous threads" necessary for the cork tree's survival. In addition, Hooke speculated—as did many of his colleagues at the time—that only plants possessed cells. Apparently it had not yet occurred to anyone to look at animal cells through a microscope.

Hooke included drawings of the cells he had observed in his book *Micrographia*. He also provided instructions for constructing a microscope like the one he used, presumably so that readers could make the same observations. Not content just to observe cells, Hooke calculated how many might be contained in a cubic inch. His accurate result: 1,259,712,000[2].

>> **FACT**: Human cells vary widely in their life spans, from 13 days to a human lifetime.

# 46 Theory of Circulation (1628)

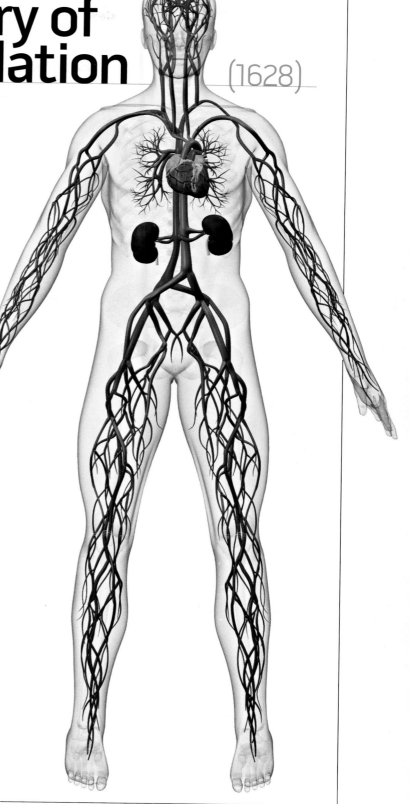

**M**uch as Copernicus broke the news that the sun, not Earth, is the center of the universe, English physician William Harvey had the task of convincing his colleagues that the beating heart—not the liver—is the fulcrum of blood circulation.

In so doing Harvey overturned understandings that had persisted for more than a millennium. The prevailing belief at the dawn of the 17th century was that the liver converted food into "natural" blood, which circulated through veins and the heart to all the body's tissues, where it was consumed. The arterial system, on the other hand, carried air—breath—and was separate from the venous circulation, though it was supplied with a small amount of blood via holes in the septum separating the heart's chambers.

Harvey's careful deductions and experiments proved this could not be so. A cut artery spurted blood in concert with the heartbeat. In a live dissection of a snake, a pinched artery seemed to engorge the heart, whereas a pinched vein made the heart shrink and go pale. Harvey furthermore calculated that the amount of blood entering the arterial system from the heart in a single hour equaled a multiple of the person's entire blood volume—far more than could plausibly be synthesized in the liver.

In 1628 Harvey published his famous treatise, saying that the movement of blood through the body is circular, and the pulsing heart drives the blood's perpetual motion.

**Harvey revolutionized medicine by writing that the heart pumps blood through arteries and veins.**

# 47 Microscope (1590)

Perhaps science owes the invention of the compound microscope in late 16th-century Holland to another innovation developed in medieval Italy: eyeglasses. It was eyeglass maker Zacharias Janssen and his father, Hans, who around 1590 discovered that they could enhance magnification by using two lenses.

People had been using single lenses, such as magnifying glasses, for centuries to examine minuscule objects. Such a lens can be made into a simple microscope, but the device is limited in its magnifying power. The compound lens structure produced by the Dutch inventors allowed an image that was magnified by one lens to be further magnified by a second.

Soon afterward, Galileo would explore the heavens with newly developed telescopes, but it would be more than a half century before microscopes would reveal a secret terrain closer at hand.

In the late 1600s Dutchman Anton van Leeuwenhoek improved the technology by creating a scope capable of magnifying objects nearly 270 times.

Today's compound microscopes typically have multiple, interchangeable lenses to give the user a choice of magnification power.

A worker uses a microscope to inspect computer chips.

>> **FACT**: Today, the strongest microscope has a half-angstrom (half a ten-billionth of a meter) resolution.

# 48 Anatomy Text (1543)

A natomist Andreas Vesalius's study of anatomy began with his determination to recover the ancient wisdom of Hippocrates and Galen, an ethnic Greek born in A.D. 129, from medieval vulgarizations. But before long Vesalius's humanist approach led him to correct some of Galen's errors via the evidence of his own senses. Like his forerunner Leonardo da Vinci, who performed human dissections in the service of science and art, Vesalius took every opportunity as a professor in Padua and later Bologna to work with cadavers, often filched from graveyards or gallows.

Vesalius was only 30 years old at the first publication of his book *On the Fabric of the Human Body*, a text in seven volumes based on his lectures in Padua.

Richly illustrated with engravings from the workshop of celebrated Italian painter Titian, Vesalius's book is an exemplar of Renaissance flowering in scientific learning, artistic technique, and printing acumen. It contains intricate and often accurate renderings of the bones (with labels in Latin, Greek, and Hebrew), as well as of the musculature, vasculature, nervous system, and urinary and reproductive tracts. The book is as much an artistic work as a clinical text. Drawings show cadavers or skeletons posed in front of pastoral or village landscapes, often in postures of considerable pathos; one skeleton is shown with its face turned toward the heavens as it leans on a shovel and gestures toward a freshly dug grave.

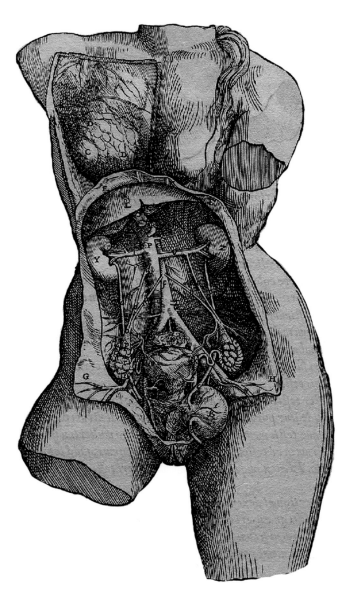

The 1543 text *On the Fabric of the Human Body* by Andreas Vesalius is considered a groundbreaking work on anatomy. The illustrations were based on dissections performed by the author (above).

# 49 Experimental Medicine (1025)

Ibn Sina, often called by his Latinized name Avicenna, was a Persian philosopher-physician who penned perhaps the most famous medical work of medieval Islamic societies, *The Canon of Medicine.* The second book of this five-volume series presents some 760 drugs that Ibn Sina judged useful, along with standards for determining efficacy that laid out the basic tenets of experimental medicine:

- Each drug, unadulterated and unspoiled, should be tested in patients with a single condition.
- The investigator should begin with the smallest dose.

- Efficacious drugs should have a consistent effect.

Ibn Sina believed in investigating the roots of illness. In the *Canon* he wrote, "The knowledge of anything, since all things have causes, is not acquired or complete unless it is known by its causes. Therefore in medicine we ought to know the causes of sickness and health."

The 14-volume *Canon of Medicine* was written by Persian scientist and physician Ibn Sina (above right).

# 50 Scientific Study of Medicine (400 B.C.)

Many scholars trace the origins of Western medicine to a radical moment in ancient Greek society, when Hippocrates and his followers first sundered the art of doctoring from that of the magician or priest.

Hippocrates, born on the Greek island of Kos about 460 B.C., may or may not have authored the early Greek writings of the Hippocratic Corpus. But it is clear from these texts that some physicians of the period were beginning to insist that disease be attributed to natural or material causes—to a patient's environment, diet, or daily habits, for example—and not to divine intervention. One text, *On the Sacred Disease,* specifically excludes mystical influence in epilepsy (which the Greeks had set apart as a strange affliction certain to reflect some divine curse or power) and instead calls it a disease of the brain. For the first time, the Hippocratic physicians located all thoughts and feelings in the brain. They counseled careful and close observation of individual patients, as well as a gentle, conservative treatment approach that strove to assist the body's own restorative powers—probably a blessing in an era of few bona fide cures.

This movement also espoused a scrupulous professionalism on the part of doctors, a value encoded in the famous Hippocratic oath and in other texts. According to Hippocratic writings, doctors should seek to benefit and never to harm their patients. They should be honest and well kept, protect the privacy of houses they enter, and avoid any form of corruption, including sexual relations with clients.

The title page from a 1559 edition of Hippocrates' complete works. The Greek physician is considered the father of medicine.

# 03

# INVISIBLE FORCES

The quest to understand the universe and humankind's place within in it scientifically goes back to the ancient Greeks. By the 16th century, it had become increasingly apparent that the universe is governed by physical laws that can be described by mathematical equations. These rules were defined clearly when Newton formulated the laws of motion and universal gravitation in the 17th century.

Just as researchers seemed to be closing in on a complete scientific understanding of nature, however, theories of 20th-century physicists—most notably Einstein's theories of relativity and theories of quantum mechanics—overturned much of what scientists thought they knew about the fundamental workings of the universe. The standard model of particle physics unites the fundamental forces of nature—with the significant exception of gravity—into a single model. Today, efforts are under way to unite quantum mechanics and general relativity into a theory of quantum gravity.

Meanwhile, breakthrough ideas in the history of chemistry led not only to an increasing understanding of the natural world, but also to the discovery and engineering of new materials with desirable properties. In 1911 Dutch physicist Heike Kamerlingh Onnes discovered the phenomenon of superconductivity when he cooled mercury to a temperature near absolute zero, the point at which no more heat can be removed from a system. Onnes's discovery ignited the continuing quest for high-temperature superconductors.

Chemistry and physics together form the background for practical innovations in transportation as well. In about 5,500 years—a relatively short period of time in the history of humankind—human beings went from inventing the wheel to traveling in electric cars and even taking flight.

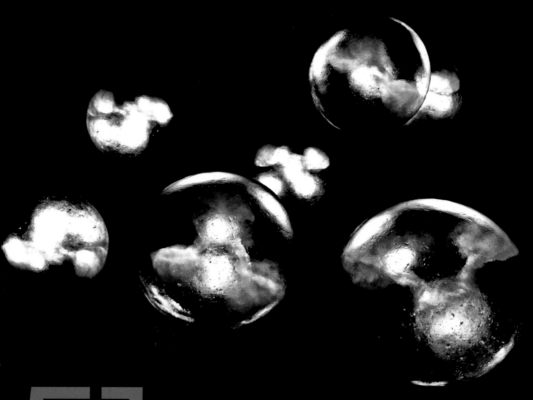

Quarks, whose existence was confirmed in the mid-1970s, are elementary particles that help make up all matter.

# 51 Standard Model of Particle Physics (1973)

The standard model of particle physics describes the universe in terms of matter, which is made up of particles called fermions, and force, which is made up of particles called bosons. There are four known forces of nature, each mediated by a fundamental boson particle: strong nuclear force, weak nuclear force, electromagnetic force, and gravity.

The standard model as it currently exists was finalized in the mid-1970s with the confirmation of the existence of quarks. The discoveries of the bottom quark, the top quark, and the tau neutrino have given this model even more credibility. Because scientists can use the standard model to explain a wide range of experimental results, it is sometimes called the theory of everything. However, this model is not without its limitations. For example, it does not take the physics of general relativity, such as gravitation and dark energy, into account.

Despite these limitations, the standard model of particle physics is one of the best explanations scientists have for how the universe works.

>> **FACT**: One elementary particle predicted by the standard model has yet to be observed: the Higgs boson.

# 52 Hybrid Vehicle (1972)

Some people may think the Toyota Prius was the first hybrid vehicle to run on both gas and electricity. Not so. That honor belongs to the hybrid Buick Skylark, which Victor Wouk conceived in 1972. The Environmental Protection Agency, apparently believing the technology was not practical, refused to approve the vehicle even though it met the clean air emissions standards of the time.

Wouk began his work on this venture in 1962, when Russell Feldman, one of the founders of Motorola, asked him to explore the possibilities presented by a solely electric car. After many studies and tests, Wouk concluded that a purely electric car was not a viable commercial venture. (In the 1970s that was likely true.) Instead Wouk proposed a hybrid vehicle that combined the advantages of an electric vehicle—namely, low emissions—with the power of a gasoline-driven vehicle.

>> **FACT:** Victor Wouk's brother, novelist Herman Wouk, supposedly based one of his characters on Victor.

A wire-frame model of a hybrid car displays aerodynamic lines.

# 53 String Theory (1969)

String theory, which was first explored by physicists who were studying the dual resonance model—a physical theory of the strong nuclear force—is an attempt to bring together quantum mechanics and general relativity by removing the discrepancies that exist between the two theories. According to string theory, electrons and quarks within an atom are not zero-dimensional objects. Instead they are one-dimensional, oscillating lines called strings. How a string vibrates determines the amount of energy that is produced and results in a specific type of subatomic particle.

Five major string theories have been formulated since the theory's beginnings in the late 1960s. In the mid-1990s these five theories were unified into what is called M-theory, which asserts that strings are really one-dimensional slices of a two-dimensional membrane vibrating in 11-dimensional space.

String theory is difficult to test. However, particle accelerators could be on the verge of finding evidence for high-energy supersymmetry—a key prediction of string theory—in the next decade.

>> **FACT:** Two basic types of string theories exist: those with closed loops that can break, and those that can't.

An illustration conceptualizes the extra dimension suggested by string theory.

A crushed plastic container is just one of the innumerable items made with polyethylene.

# 54 Polyethylene (1954)

It is hard to imagine daily life without polyethylene. From plastic grocery bags to laundry baskets and gallon milk containers, polyethylene products are everywhere. The production of polyethylene is now a $7 billion industry, but it all started by accident in 1953.

German chemist Karl Ziegler was working at the Kaiser Wilhelm Institute (later named the Max Planck Institute). He analyzed free radicals—molecules that have an uneven number of electrons, and thus look to bond with other molecules—and the reactions of ethylene. During an experiment, a trace amount of nickel was accidentally left in the chamber of Ziegler's apparatus. The result was a long chain of carbon atoms that had not been seen before. Ziegler conducted similar experiments with other metals and found that aluminum produced the best results: a plastic structure that was not only strong but also bendable. Although plastics had been created before this, Ziegler's process was considered a breakthrough because it could be done at close to room temperature and atmospheric pressures, rather than the very high pressures and temperatures that other methods required.

Giulio Natta, an Italian chemist, expanded on Ziegler's work and used his process to develop polypropylene, a material that is used in products such as food containers and carpet fibers. Natta and Ziegler were jointly awarded the Nobel Prize in chemistry in 1963.

Today, people have voiced concerns about plastic's lack of biodegradability, thus leading to research into biodegradable forms of plastic.

# 55 Quantum Mechanics (1927)

Quantum mechanics is a field of study dedicated to the behavior of matter on atomic and subatomic scales. It came about in the 1920s in response to the realization that classical physics could not explain certain phenomena, something first noted in physicist Albert Einstein's explanation of the photoelectric effect, also known as the general theory of relativity. The principles behind quantum mechanics can be hard to grasp, especially since scientists are used to viewing the world through the lens of classical physics.

A few of the main aspects of the theory that capture its weirdness are wave-particle duality, the uncertainty principle, and superposition—the last of which plays a major role in quantum computing. Wave-particle duality refers to the fact that on small scales light and matter have both wave- and particle-like properties. The uncertainty principle says that both a particle's position and its momentum cannot be known with certainty—the more accurately one of these quantities is measured, the less accurately the other is known. Superposition is the quantum mechanical idea that a particle can exist in all possible states at the same time. As odd as the theories of quantum mechanics seem, physicists have determined through experiments that the theories do accurately represent the nature of reality on very small scales.

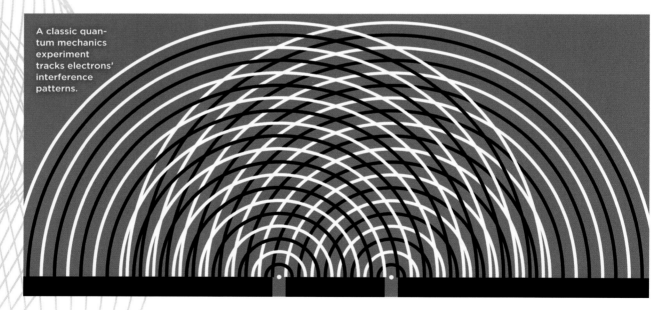

A classic quantum mechanics experiment tracks electrons' interference patterns.

>> **FACT**: Quantum mechanics has made possible the theoretical development of quantum computers.

# 56 General Theory of Relativity (1915)

When Albert Einstein introduced his general theory of relativity in 1915, it showed that Newton's law of universal gravitation—which had been accepted for more than 200 years, and which formed the foundation of how scientists understood the universe—was only partially correct. It no longer applied when the gravitational force became too strong. While Newton's laws are very accurate at explaining most kinds of motion, they cannot predict the exact behavior of very massive objects, such as planets in orbit.

With Einstein's earlier special theory of relativity, time was no longer objective and absolute, and space and time could be considered to be united in the single four-dimensional continuum of space-time. General relativity was a completely new, astounding mathematical theory that interpreted the force of gravity as a curvature in space-time.

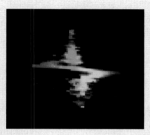

Einstein's formulation of this theory began with a simple thought experiment: What happens when someone falls out of his chair? Starting from the key insight that a person in free fall does not feel his own weight, he constructed a theory that models gravity as curved space-time. General relativity provided not only a new way to interpret gravity, but also a new framework within which to understand the evolution of the universe. For example, the mathematical aspects of the big bang theory are based on general relativity. Also, using the general theory of relativity, cosmologists and astronomers were able to predict accurately, and subsequently prove, the existence of neutron stars, black holes, and gravitational waves.

Images of a 1919 total solar eclipse helped confirm Einstein's theory of general relativity—which, in turn, aided astronomers in understanding the data from black holes such as M84 (above).

57

# Quantum Gravity

Quantum gravity is an area of physics research in which scholars are attempting to unify Albert Einstein's theory of relativity with quantum mechanics, also known as quantum theory. The difficulty of this task lies in the two theories' very different, contradictory approaches to how the universe works. There are four fundamental forces of nature: gravity, electromagnetism, strong nuclear force, and weak nuclear force. These are the ways in which individual particles interact with one another. According to classical physics, these forces have definite and specific strengths, directions, velocities, and masses, which can be used to determine the curvature of space-time. According to quantum mechanics, however, the forces do not have definite values, and they are affected by subatomic particles such as photons, W and Z bosons, gluons, and the graviton. Thus, whereas scientists might expect an atom of matter to behave in a certain way when looked at through the lens of classical physics, this same atom could behave in quite different ways when seen through the lens of quantum physics.

Quantum gravity is a theory that is unconfirmed—and is likely to remain so for the foreseeable future. For one thing, the energy levels required to observe the phenomena that the theory predicts cannot be achieved in current laboratory experiments. In addition, Einstein's theory of general relativity makes certain assumptions about the universe at the macroscopic scale that are quite different from the assumptions of quantum mechanics, which works at the microscopic scale.

Scholars are conducting several different lines of research into quantum gravity. But many researchers believe that even more radical concepts of space-time must be formulated before quantum gravity can be properly studied. Scientists also disagree about whether quantum gravity is an entirely new field, with new concepts, or whether it is putting a new face on old mathematical differences and contradictions between general relativity and quantum mechanics.

• **51** The **STANDARD MODEL** incorporated the forces of nature, with the exception of gravity, into a single theory of particle physics.
• **53 STRING THEORY** is a significant—yet unverified—theory of quantum gravity.

• **56** The **GENERAL THEORY OF RELATIVITY** gives scientists a mathematical description of the force of gravity and how it affects time and space.
• **69 NON-EUCLIDEAN GEOMETRIES** provide new ways to model space and time mathematically.

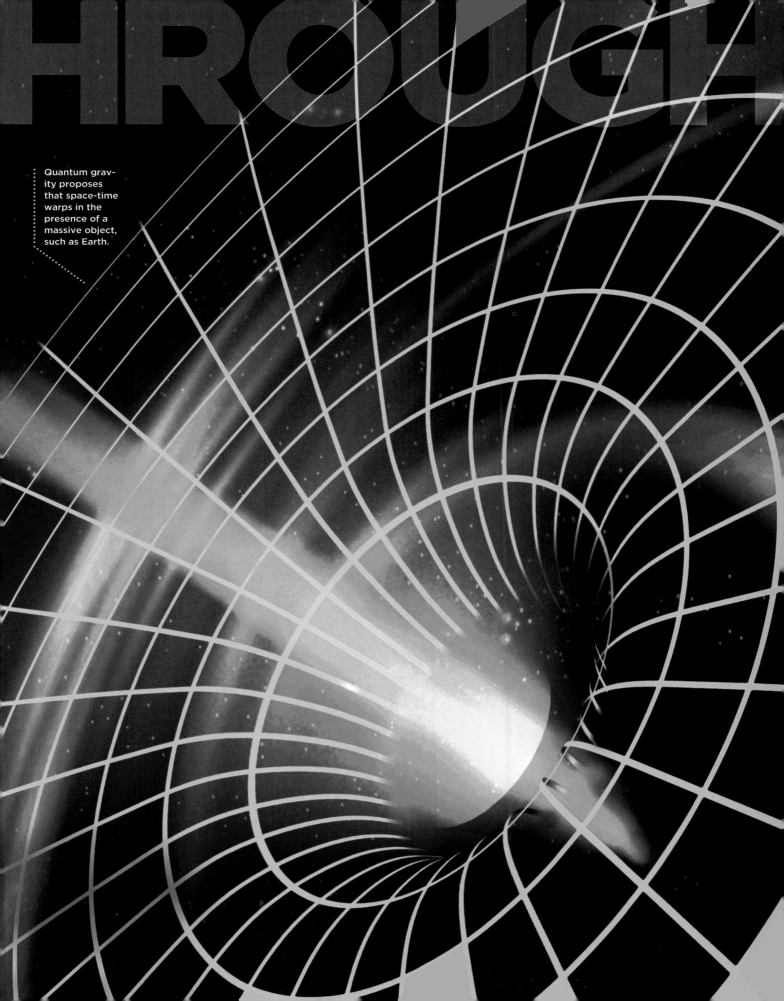

Quantum grav-
ity proposes
that space-time
warps in the
presence of a
massive object,
such as Earth.

# 58 Bohr Model of the Atom (1913)

When Danish physicist Niels Bohr joined forces with British physicist Ernest Rutherford at Manchester University, his primary goal was to improve on Rutherford's model of the atom, which had been introduced in 1911. Rutherford's model depicted the atom with electrons orbiting the nucleus, in much the way that a solar system's planets orbit the sun. The model was flawed, however: The atom would have much too short a life span because the electrons would lose energy, emit electromagnetic radiation, and spiral inward in an unstable fashion.

While analyzing a hydrogen atom, Bohr quickly found the solution to Rutherford's problem and created his own model of the atom in 1913. Bohr's model was the first to use quantum mechanics to describe the behavior of an atom. In this model, when an electron absorbs electromagnetic radiation, the electron jumps into a different—but very specific—orbit. That is, the electrons can only occupy the orbits prescribed by Bohr's theory; they cannot be in between orbits. Because there is an innermost allowed orbit, the electrons do not spiral into the nucleus.

Using his model, Bohr was able to calculate the energies of the orbits of hydrogen and similar atoms. Bohr's model advanced the study of atoms, and he was awarded the Nobel Prize in physics in 1922.

While Bohr's model is still used to introduce the topic of atomic structure and behavior to new physics students, it is not considered the best model. The model used today is called the electron cloud model, which depicts electrons in a random cloud pattern around the nucleus, rather than orbiting it.

The classic model of a hydrogen atom proposed by Niels Bohr (above) is based on a positively charged nucleus orbited by electrons (below).

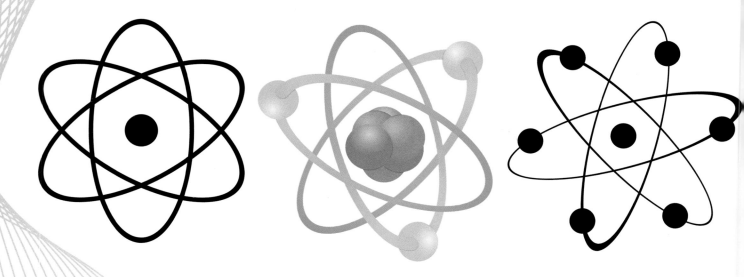

A transverse section of a superconductor

# 59 Superconductivity (1911)

n the early 1900s, scientists including Heike Kamerlingh Onnes were studying low-temperature refrigeration. Onnes wanted to investigate the effects of low temperatures on the properties of metals. There were several theories about what would happen to electrical current passed through a metal cooled to absolute zero (-459.67°F). Some scientists thought the metal's resistance to electricity would diminish, while others thought it would increase.

Onnes discovered that when mercury was cooled to a temperature close to absolute zero, its resistance to electricity disappeared altogether. That is, the electrical current that he passed through the mercury did not diminish or dissipate. He called this phenomenon superconductivity.

In the years that followed Onnes's discovery, other materials—including tin, aluminum, and lead—were found to be superconductors when cooled to very low temperatures. In 1986 physicists Karl Alexander Müller and Johannes Georg Bednorz discovered a ceramic composite that superconducted at a higher temperature, near 30K (-405.4°F). If scientists could create a superconductor at room temperature, they could create electricity that would flow unceasingly.

Potential applications of superconductors have already begun. In Japan engineers have used superconducting magnets to construct a prototype of a levitated train. Magnetic resonance imaging, or MRI, a technology that also makes use of superconducting magnets, continues to play a very important role in diagnostic medicine.

>> **FACT:** Superconducting wires have been known to carry electricity for years with no measurable loss.

In the early 20th century, physicist Albert Einstein revolutionized physics with his special theory of relativity.

# 60 Special Theory Of Relativity (1905)

I n 1905 Albert Einstein published the special theory of relativity, which made use of two key physical ideas that were known previously: the principle of relativity and the constant speed of light.

Prior to the theory of relativity, physicists believed that electromagnetic waves moved through a medium called ether, much like ocean waves move through water. They viewed ether as a background against which all movement took place; all objects in motion moved relative to the ether. Einstein felt that it was a mistake to assume the existence of ether, which had not been verified experimentally. In the construct of special relativity, he did away with ether altogether, assuming only that the laws of physics, including the speed of light, worked the same no matter how an observer was moving.

The mathematical consequences of Einstein's theory were stunning, and they have been experimentally verified. For example, as an object moves with a velocity relative to an observer, the object's mass increases and its length contracts. Perhaps the most famous consequence of the theory is the equivalence of mass and energy that is captured in the equation $E = mc^2$. The special theory of relativity also created a fundamental link between space and time, a four-dimensional framework called the space-time continuum. This continuum consists of three dimensions representing space—up/down, left/right, forward/backward—and one dimension representing time.

Einstein's theory is considered special because it applies the principle of relativity only to the special case in which the motion of objects is uniform.

# 61 First Motor-Powered Aircraft (1903)

On December 17, 1903, Orville and Wilbur Wright proved not once but four times that sustained, motor-driven air flight was possible. Orville made the first flight, which covered a distance of 120 feet and lasted 12 seconds. Wilbur made the final one, which covered an impressive 852 feet and lasted 59 seconds.

The Wright brothers had prepared for this moment for a long time. They used gliders to perfect their technique and to gain a better understanding of how to control the "official" craft, called the *Wright Flyer.*

They also spent a lot of time watching birds in flight. Their observations showed them that it was the curve of the birds' wings that gave them lift. In addition, birds maneuvered by changing the shapes of their wings. Thus Orville and Wilbur knew that they would need to employ the same technique, known as wing warping. Wing warping enables a pilot to keep a craft steady by arching the wing tips slightly.

When they realized that an automobile engine was going to be too heavy for the craft, the brothers designed and built their own.

>> **FACT**: Tip to tip, the wingspan of the *Wright Flyer* measured 12.3 meters (40.3 ft).

Orville and Wilbur Wright make the first powered flight over the sands of Kitty Hawk, North Carolina.

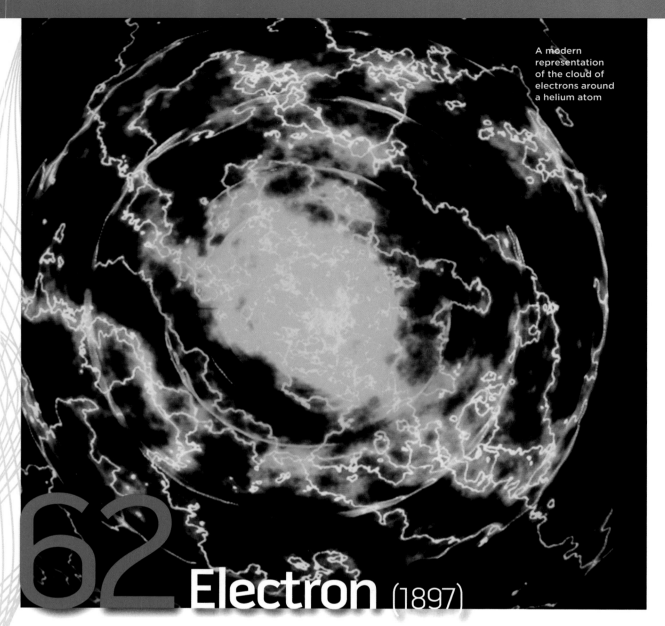

A modern representation of the cloud of electrons around a helium atom

# 62 Electron (1897)

English scientist John Dalton introduced the theory of the existence of atoms in the early 1800s. Sir Joseph John "J. J." Thomson, a physicist, wanted to expand on Dalton's theory. Most of Thomson's work consisted of conducting electricity through gases, which he did in the study of cathode rays (streams of electrons in vacuum tubes). In 1897 Thomson realized that the cathode ray was actually a stream of electrons given off by an atom, and this led him to formulate the theory that because an atom is neutral, it consists of an equal number of negatively and positively charged particles. This was later called the plum pudding theory of the atom, because when drawn as a diagram, the atom resembled a British plum pudding—the atom is the batter, and the charged particles are the pieces of fruit.

Just over a decade later, the work of physicists Ernest Rutherford and Niels Bohr furthered the development of atomic theory. Rutherford proposed that the atom had a nucleus with electrons orbiting it, much the way a solar system's planets orbit the sun, but this theory was flawed because it made for an unstable atom. Bohr solved this problem by using quantum mechanics to describe the behavior of an atom.

# 63 Speed of Light (1887)

n the late 19th century physicists believed that just as sound and other substances must have a medium to move through—such as water or air—so, too, must light. This medium was called ether, and several scientists undertook the task of proving its existence.

The most famous experiment was performed by American scientists Albert Michelson and Edward Morley in 1887. They constructed an elaborate apparatus, known as an interferometer, that would allow them to detect ether. Basically, they would measure the speed of light in two directions, one the same as Earth's motion, the other perpendicular to Earth's motion. If ether existed, the beam of light traveling parallel to Earth's motion should be slower—that is, it should take longer to travel an equal distance—than the beam traveling perpendicular to Earth's motion. This,

they hoped, would establish the existence of the ether.

Michelson and Morley expected to find different speeds of light, but they found no discernible changes in speed. Thus, the ether did not exist.

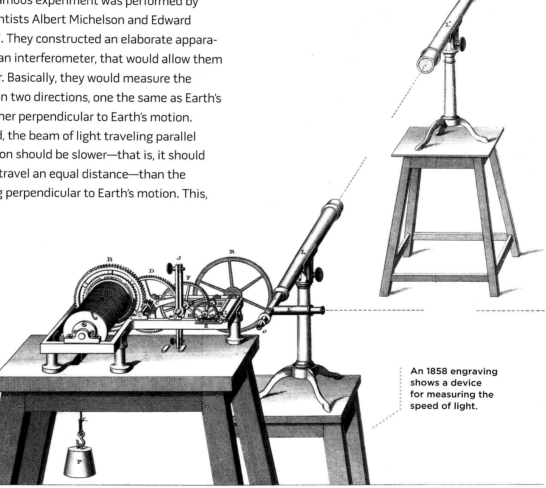

An 1858 engraving shows a device for measuring the speed of light.

>>**FACT:** Light travels at a fixed speed of 186,000 miles (299,338 km) per second.

# 64 Room-Temperature Superconductivity

The potential applications of room-temperature superconductivity have captured the imaginations of scientists and resulted in an abundance of research in the last 100 years. Superconductors are materials that conduct electricity with zero loss of energy or resistance. Most superconductive materials discovered to date must be cooled to hundreds of degrees below freezing in order to attain superconductivity. If scientists could discover a material and method for creating room-temperature superconductivity, it would be a world-changing breakthrough.

Because using superconductive materials in everyday life would mean transmitting power with zero loss of energy, tremendous amounts of energy could be saved. The realization of room-temperature superconductivity could also lead to superfast elevated trains, superefficient magnetic resonance imaging (MRI), powerful supercomputers, superconducting magnetic energy storage, and many other potentially life-altering inventions. In fact, many people have called room-temperature superconductivity the holy grail because of the vast number of important applications that could result from its discovery.

Superconductivity was discovered accidentally in 1911 when Dutch physicist Heike Kamerlingh Onnes

High-speed trains in Germany (shown), Japan, and China use mag-lev—magnetic levitation—to transcend friction and increase speed.

So far, only in lab settings (shown) can superconductors generate enough magnetic force to levitate objects at room temperature.

and his assistant Gilles Holst noted that mercury's conductivity resistance dropped to zero at temperatures below 4.2K (-452.1°F). In the last 30 years researchers have worked to create superconductors at higher temperatures. In 1986 physicists Karl Alexander Müller and Johannes Bednorz were able to achieve superconductivity in lanthanum barium copper oxide—considered the first high-temperature superconductor—at a temperature of 35K (-396.67°F). In 2001 Japanese scientists Jun Nagamatsu, Norimasa Nakagawa, Takahiro Muranaka, Yuji Zenitani, and Jun Akimitsu discovered that magnesium diboride becomes a superconductor

at 39K (-389.47°F). In 2006 researchers discovered pnictides, a group of iron-based compounds that become superconductive at 50K (-369.67°F). Scientists have yet to identify a material that is superconductive at room temperature—but the quest continues.

• **59** The discovery of **SUPERCONDUCTIVITY** in 1911 was a watershed moment on the way to room-temperature superconductivity.

• **68** Lord Kelvin's formulation of the concept of **ABSOLUTE ZERO** paved the way for low-temperature technologies like superconductors.

# 65 Automobile (1885)

In 1885 German engineer Karl Benz introduced the first gasoline-powered automobile and forever changed the world of transportation. Others had worked on the same idea independently, but Benz was the first to patent it. The automobile used an internal combustion engine, had three wheels—Benz built a four-wheeled car in 1891—and was steered using a tiller, much like a boat is steered. The engine had a four-stroke design in which the first stroke drew in an air-fuel mixture, the second stroke compressed it, the third stroke ignited it, and the fourth stroke exhausted the combustive force.

The vehicle had its shortcomings, to be sure: It was somewhat hard to control, and it had no gears, so it could not climb hills on its own power. Gasoline was only available in small quantities from pharmacists; in those days it was used as a cleaning product.

Benz had a hard time getting anyone to pay significant attention to his invention. It was not until his wife and children took the car one morning and drove it on a 65-mile trip from Mannheim to Pforzheim as a publicity stunt that people started to take an interest. Benz and Company became the largest manufacturer of automobiles by 1900, and around this time it merged with Gottlieb Daimler's company, which was producing four-wheeled vehicles. The company is known today as Mercedes-Benz. The Mercedes part of the name comes from one of Daimler's business partners, Emil Jellinek, who would invest capital in the company only if the car were named after his daughter.

The driver of this 1898 Packard directs it with a tiller instead of a steering wheel.

A melting ice cube illustrates the second law of thermodynamics.

# 66 Laws of Thermodynamics (1870)

The laws of thermodynamics were developed in the 1800s to explain the absence of perpetual motion in nature. They are expressed in three basic tenets. The first law has to do with the conservation of energy. It states that energy in a system can be transformed, but it cannot be created or destroyed.

The second law of thermodynamics states that heat cannot spontaneously flow from a colder location to a hotter location. Over time, differences in temperature, pressure, and chemical potential tend to even out in an isolated physical system, and this evening out can be measured in terms of what is called entropy.

The third law of thermodynamics states that entropy is dependent on temperature, which provides an absolute reference point for its determination. In other words, as temperature approaches absolute zero, a system's entropy approaches a constant minimum and all processes end.

Together these laws have become some of the most important fundamental laws in physics.

>> FACT: The second law says that the universe will eventually reach a state of heat death.

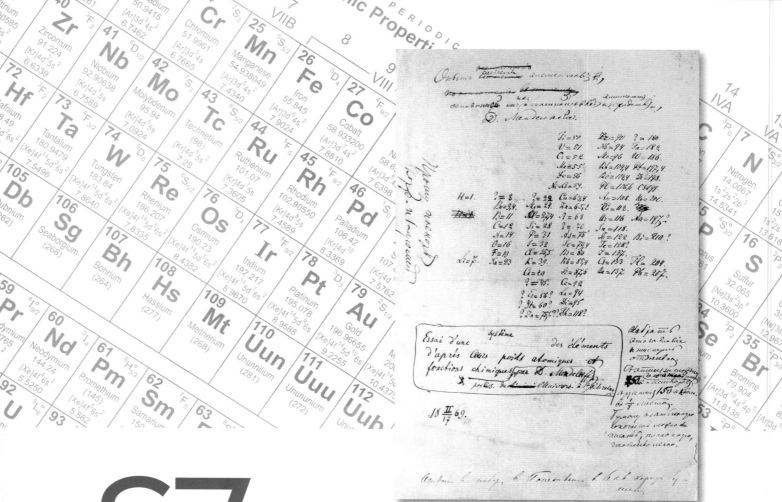

# 67 Periodic Table (1869)

Dmitri Mendeleev, a Russian chemist, created the first periodic table by using individual cards that he laid out like a game of solitaire. He made a card for each of the 63 elements known at the time; the cards included the elements' atomic weight and inherent properties. He placed each card on a table in order by ascending weight, creating columns and rows that indicated relationships and similarities in properties. Mendeleev's periodic table was included in his 1869 work *On the Relationship of the Properties of the Elements to Their Atomic Weights.*

While Mendeleev was the first to create a table this extensive, the work of several other scientists preceded his periodic table. In 1864 John Newlands, a chemist in England, created a table that grouped the 56 known elements at the time. Lothar Meyer, of Germany, created a table similar to Mendeleev's at around the same time, but it included only 28 elements.

Mendeleev's table was considered most significant because he left spaces for elements that had not been discovered yet.

**Russian chemist Dmitri Mendeleev's periodic table of 1869 rests atop a modern version.**

# 68 Absolute Zero (1848)

Even though Irishman Robert Boyle had first proposed the idea of an absolute coldest temperature in 1665, it was not until 1848, when Lord Kelvin actually devised a scale based on the idea, that the concept of absolute zero became significant.

Kelvin—William Thomson, Baron Kelvin of Largs—was a Scottish physicist and engineer whose work on heat, electricity, and the unified nature of matter and energy presaged later revolutions in atomic physics. Kelvin saw the need for an absolute temperature scale in which the lowest possible temperature would be set at zero. The freezing point of water—or more properly, the "triple point" at which solid, liquid, and gaseous water can exist simultaneously—became 273.16 kelvins, equivalent to 0°C or 32°F.

The kelvin scale became the standard for scientific measurement, particularly at the high end of temperatures, such as those in stars, and at the low end, where motion slows to an atomic crawl.

As an object cools, the molecules that make up the object move more slowly. At absolute zero, this movement is the slowest possible—and some scientists argue that motion stops altogether. This temperature is zero on the kelvin scale, which corresponds to -273.15° on the Celsius scale and -459.67° on the Fahrenheit scale.

A pair of mounts for air thermometers used in an absolute zero experiment

>>**FACT:** A Bose-Einstein condensate is a state of matter consisting of a particular gas near absolute zero.

# 69 Non-Euclidean Geometries (ca 1830)

Euclidean geometry, which dates to 300 B.C., makes use of five axioms, or postulates—statements that are taken as true. These postulates are used to prove many theorems about geometric figures and other mathematical objects. Euclid's fifth postulate, also known as the parallel postulate, is as follows: Given a straight line and a point $p$ that is not on that line, there is exactly one straight line through $p$ that never intersects the original line.

In the 1800s mathematicians discovered that they could modify the fifth postulate and thus create very different—yet mathematically consistent—geometries. The discovery of non-Euclidean geometry had wide-reaching consequences, especially in physics, because it showed that conceptions of space and time other than the familiar Euclidean one were possible.

The two primary types of non-Euclidean geometry are elliptical geometry and hyperbolic geometry. In elliptical geometry, given a line, one cannot find a line that is parallel to that line. In hyperbolic geometry, the Euclidean parallel postulate also does not hold.

>>**FACT:** Some people view elements of *Alice in Wonderland* as a satire of non-Euclidean geometries.

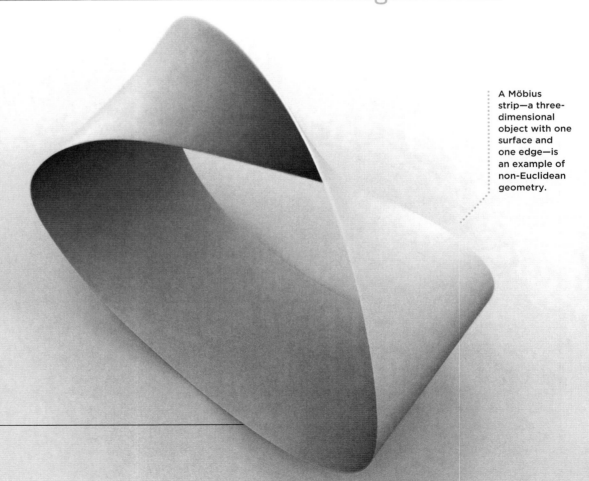

A Möbius strip—a three-dimensional object with one surface and one edge—is an example of non-Euclidean geometry.

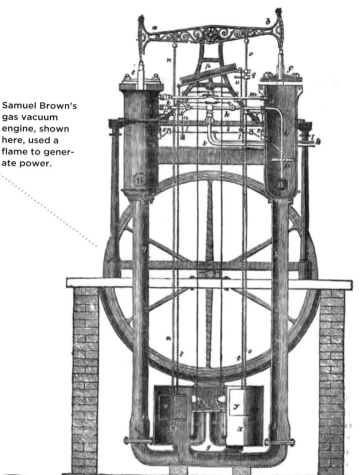

Samuel Brown's gas vacuum engine, shown here, used a flame to generate power.

# 70 Internal Combustion Engine (1823)

I n 1823 English inventor and engineer Samuel Brown developed an internal combustion engine based on an earlier steam engine model. While not the first of its kind, Brown's engine was the first to be used in an industrial capacity: to pump water and to propel boats and barges on rivers. Brown's engine, a modified version of the Newcomen steam engine, used hydrogen for fuel instead of steam and had separate compartments for combustion and cooling; it also used cylinders. Earlier internal combustion engines, such as those designed by French inventor François Isaac de Rivaz and Dutch scientist Christiaan Huygens, used different fuels and slightly different designs, and none had the same degree of effectiveness and success as Brown's combustion engine.

## >> FACT: The internal combustion engine was integral to the development of the automobile in 1885.

The Tesla Roadster's microprocessor-controlled lithium-ion battery has a 3.5-hour charge time, with an expected life span of seven years or 100,000 miles.

# 71 Electric Car

Many people might think of the electric car as a new innovation in transportation. However, electric cars were quite popular between the mid-19th and early 20th centuries because they offered a level of comfort and ease of operation that gasoline cars of the time could not provide.

Although electric cars have the potential to reduce pollution significantly by way of zero tailpipe emissions, they are not necessarily pollution free. Electricity is required to run these vehicles, and carbon dioxide is still released somewhere if the electricity comes from fossil fuels. How much carbon dioxide is released depends on the power source used to charge the vehicle (the stronger the power, the greater the emissions), how efficient the vehicle is, and how much energy is given off as waste when the car is charging.

There are currently two types of electric cars: one powered by a gasoline generator (called a hybrid) and one powered by an onboard battery pack (considered a true electric car). Today, there is no shortage of prototype, preproduction, and concept electric cars

BREAKT

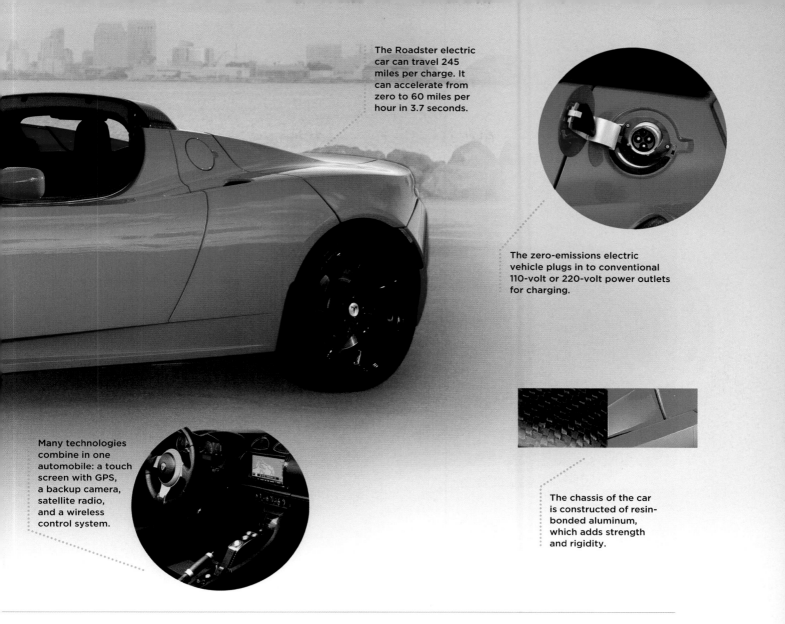

The Roadster electric car can travel 245 miles per charge. It can accelerate from zero to 60 miles per hour in 3.7 seconds.

The zero-emissions electric vehicle plugs in to conventional 110-volt or 220-volt power outlets for charging.

Many technologies combine in one automobile: a touch screen with GPS, a backup camera, satellite radio, and a wireless control system.

The chassis of the car is constructed of resin-bonded aluminum, which adds strength and rigidity.

out there; however, only a few highway-capable models are on the market. The rest are vehicles capable of only low speeds and limited ranges.

What is keeping the electric car from being widely adopted as a mode of transportation? Factors include the costs of developing, producing, and operating electric vehicles compared to those of internal combustion engine vehicles, and the fact that electric cars therefore have a higher price tag than their gasoline-powered counterparts. Another factor is what experts dubbed range anxiety. Manufacturers have

marketed these vehicles as ideal for short trips of around 40 miles. Longer trips would require some sort of switching technology or rapid recharge system, both of which are being investigated.

• **65** The invention of the **AUTOMOBILE** was a significant step on the way to the electric car.
• **73** The creation of the first **ELECTRIC CELL BATTERY** was an important milestone on the way to the battery-powered electric car.

HROUGH

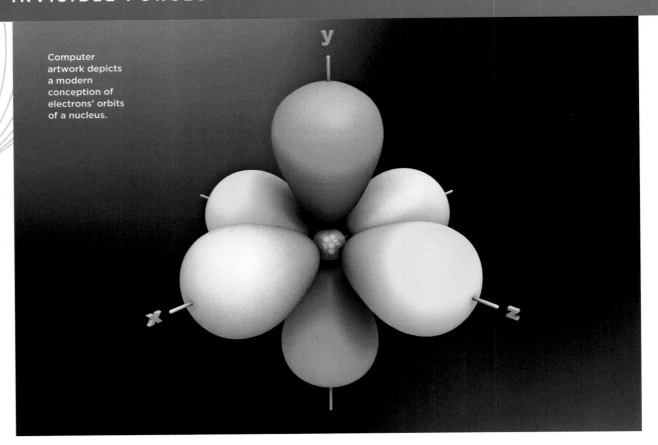

Computer artwork depicts a modern conception of electrons' orbits of a nucleus.

y

x

z

# 72 Atomic Theory (1808)

The idea that matter is made up of atoms was not a new one when English chemist and physicist John Dalton published his atomic theory in the book *New System of Chemical Philosophy*. Even the Greek philosopher Democritus, who lived in the fifth century B.C., had proposed that matter was made up of tiny component particles. What was different about Dalton's theory, however, was that it was the result of careful measurements and observations rather than philosophical debate. Although not all scientists agreed with him, Dalton's atomic theory was widely accepted by the mid-1800s. Today, he is known as the father of atomic theory.

Dalton formed his theory after years of observing the atmosphere and taking notes on how gases and liquids are formed and how they react to varying weather conditions. This led him to the first principle of his theory: Matter is composed of atoms that cannot be divided or destroyed. His second principle is that all atoms of a specific element have the same mass and properties. He also proposed that chemical compounds are formed when two or more different atoms combine and that atoms are rearranged as a result of chemical reactions. In *New System of Chemical Philosophy*, Dalton provides a list of elements along with their atomic weights, again based on his measurements and observations.

# 73 Electric Cell Battery (1800)

When Italian physicist Alessandro Volta read about the experiments of Italian professor Luigi Galvani, he was inspired. Galvani had discovered that if he touched the nerve of a frog with an electrostatically charged plate, it caused a muscle contraction. He theorized that the frog contained a certain amount of electricity. Volta set out to prove Galvani wrong. Volta felt it was the combination of the metal and water that had created the charge.

In 1800, to test his hypothesis, Volta created what is now known as a voltaic pile. The pile consisted of two stacks of metal—alternating copper and zinc, with cardboard soaked in salt water sandwiched between each metal piece. The piles created an electrical charge; when Volta connected the two piles with wires, they sustained that charge for a period of time. This was the first battery.

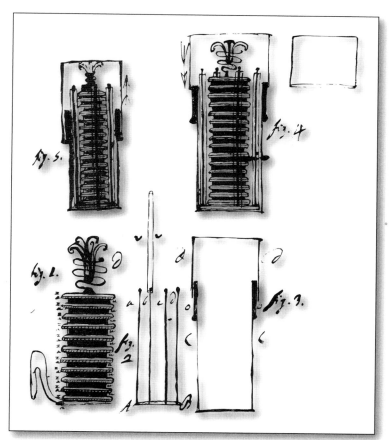

The first electric battery, shown in an early 18th-century drawing by inventor Alessandro Volta, eventually led to today's alkaline battery.

# 74 Electricity (1745)

Some of the most important discoveries paving the way for modern applications of electricity occurred in the 18th century.

In November 1745 German scientist Ewald Georg von Kleist invented a simple capacitor, which stored an electrical charge. Around the same time, Pieter van Musschenbroek, a Dutch professor at the University of Leyden, came up with a similar device in the form of the Leyden jar. The two men's work proved that static electricity could be transformed into an electric current.

German physicist Daniel Gralath combined several Leyden jars to increase the storage capacity.

In 1747 American inventor Benjamin Franklin investigated this further and proved that the charge was stored on the glass, not in the water. In addition, Franklin established the link between lightning and electricity during his famous kite-in-a-thunderstorm experiment.

Other important figures during this time include Michael Faraday, André-Marie Ampère, Georg Ohm, Luigi Galvani, and Alessandro Volta, all of whose names are attached to some measure of electricity.

**Benjamin Franklin proved that lightning was an electrical current.**

# 75 Newton's Laws of Motion and Universal Gravitation (1687)

According to Newton's law of universal gravitation, every massive particle in the universe is attracted to every other massive particle with a force that is directly related to the sum of their masses and is inversely related to the square of the distance between them. Proportionality is held constant in this theory at all places and all times; thus, it is known as the universal gravitational constant.

Newton's laws of motion consist of three physical laws that describe the relationship between forces acting on a body and its motion due to these forces. They can be summarized as follows:

(1) A body remains at rest or in uniform motion in a straight line unless acted upon by a force.

(2) The acceleration of a body is proportional to the force causing the acceleration and is inversely proportional to its mass.

(3) When a force acts on a body due to another body, an equal and opposite force acts simultaneously on that body.

Taken as a whole, these laws are significant in that they established the notion of a clockwork universe, which permeated science for the next 200 years. In this view the universe is akin to a clock wound up by God and set in motion, with everything running as a perfect machine. Deists embraced this idea and delighted in the concept that God simply set the wheels of the universe in motion. Newton was dismayed by this interpretation; he feared it would lead to atheism.

**Newton's laws allowed astronomers to explain planetary orbits.**

# 76 First Theory of Atoms (400 B.C.)

Although some people might view the atom as a relatively modern discovery, thinkers speculated upon its existence as far back as 400 B.C. The first was Greek philosopher Leucippus, followed by his student Democritus. Leucippus believed that everything is composed of various imperishable, indivisible elements called atoms (the word comes from the Greek words meaning "unable to cut"). Democritus took his mentor's work a step further by theorizing that the solidness, or lack thereof, of a given material was directly related to the shapes of its atoms. Iron atoms were solid and strong, for example, whereas air molecules were light and whirling.

>> **FACT**: Building on the work of botanist Robert Brown, Einstein proved the existence of atoms in 1905.

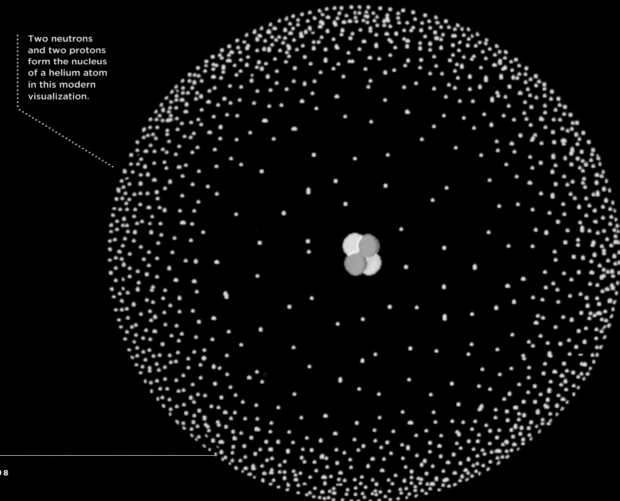

Two neutrons and two protons form the nucleus of a helium atom in this modern visualization.

Ancient people found they could move heavy objects by placing beneath them a round device attached to an axle.

# 77 Wheel (3500 B.C.)

Among the most revolutionary ideas known to humankind, the wheel may have been invented in Mesopotamia around 3500 B.C. The wheel's potential was recognized in areas of manufacturing and industry such as pottery before it was applied to transportation. Eventually wheels appeared on chariots, carts, wagons, and anything else a person might want to move. Despite the wheel's usefulness, it was not used as a means of transportation in ancient cultures in the Americas. Nor did the wheel evolve much beyond its original form until the 19th century and the advent of the industrial revolution.

# THIS WO
# AND OTH

Scientists have come a long way in their understanding of Earth—both the forces that shape it and the life it nurtures—since Aristotle first took a stab at organizing and classifying animals in the fourth century B.C. Early views of Earth as a static, unchanging rock have given way to knowledge of plate tectonics, continental drift, and climate change. Scientists know that change on this planet happens in fits and starts, and that these fits and starts tend to occur consistently over long periods of time. As early as 1896, Swedish scientist Svante Arrhenius warned that human activity was affecting Earth's climates. In the future, geoengineering may allow scientists to counteract the effects of climate change.

Understanding how natural forces shape Earth has also led us to look beyond our own planet. It is human nature to wonder: Are we alone? What more is there to discover about the universe? From the revolutionary ideas of Copernicus

RLD
ERS

to the big bang theory, astronomers have sought to provide a framework for understanding the evolution of the universe as a whole from its earliest moments. Scientists are now searching for Earth-like planets in other parts of the galaxy—a quest made possible by the Hubble Space Telescope and other powerful instruments.

Meanwhile, we are taking our first steps into our solar neighborhood. From 1961, when the Soviet Union put the first man into space, it was only eight years until the United States answered by landing men on the moon during the Apollo 11 missions. There is a long way to go, but far on the horizon is the possibility of "terraforming" a neighboring planet—perhaps Mars—with the goal of making it habitable for humans.

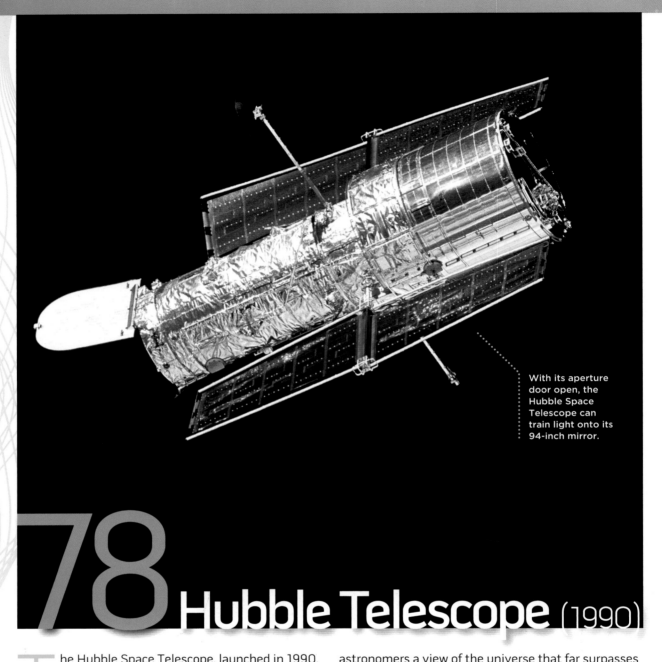

With its aperture door open, the Hubble Space Telescope can train light onto its 94-inch mirror.

# 78 Hubble Telescope (1990)

The Hubble Space Telescope, launched in 1990, is both a spacecraft and a telescope. Located 370 miles above Earth's surface, it orbits the planet once every 97 minutes. Its distance above Earth is significant. Earth's atmosphere distorts and blocks the light that reaches the surface of the planet. Thus, placing Hubble beyond the atmosphere gives astronomers a view of the universe that far surpasses that of ground-based telescopes.

As a result of Hubble's launch, astronomers hope to answer important questions about the age and size of the universe; how planets, stars, and galaxies form; and how stars are born and die.

>> **FACT**: Hubble orbits Earth once every 97 minutes, at a speed of five miles (eight km) per second.

# 79 Inflationary Theory of the Universe (1980)

Inflationary theory was developed by Alan Guth in 1980 to explain observable features of the universe. According to the theory, the universe is the result of an extremely rapid but short-lived expansion—known as the inflationary epoch—during its early history.

Precursors to inflationary theory include thinking by Albert Einstein, several Soviet cosmologists, and others, who attempted to explain why the universe appears flat, homogeneous, and uniform in all directions—rather than highly curved and heterogeneous.

The big bang, the prevailing theory of the development of the universe, is a classic example of inflationary theory at work. Astronomers have made precise measurements of the cosmic radiation left over from the big bang and have determined that the radiation arrives at Earth from all directions with the same intensity. Tracing the development of this radiation backward, cosmologists concluded that the temperature and density of matter in the universe must have been uniform when the cosmic background radiation was released.

The Helix Nebula is a remnant of a star that has formed and died since the big bang, more than 13 billion years ago.

# 80 Mass Extinction (1980)

I n 1980 Luis Alvarez, an American experimental physicist and inventor, along with his son, geologist Walter Alvarez, made a major discovery about the history of life on Earth. The two men were studying a 65-million-year-old layer of clay that was deposited during the transition from the Cretaceous period to the Tertiary period. The clay contained an unusually high amount of iridium. Luis Alvarez reasoned that the most plausible source of this element had to be extraterrestrial—in particular, a comet or an asteroid—as iridium is exceedingly rare in Earth's crust. The Alvarezes put forward what became known as the impact theory: the conjecture that the extinction of the dinosaurs was a result of a comet or asteroid directly hitting Earth.

In a 1982 paper Jack Sepkoski and David M. Raup described the Big Five mass extinctions: End Cretaceous (the one the Alvarezes studied), End Triassic, End Permian, Late Devonian, and End Ordovician. These are not, however, the only mass extinction events that have occurred in Earth's history.

Zuniceratops (below) and ammonites (above) became extinct about 90 and 65 million years ago, respectively.

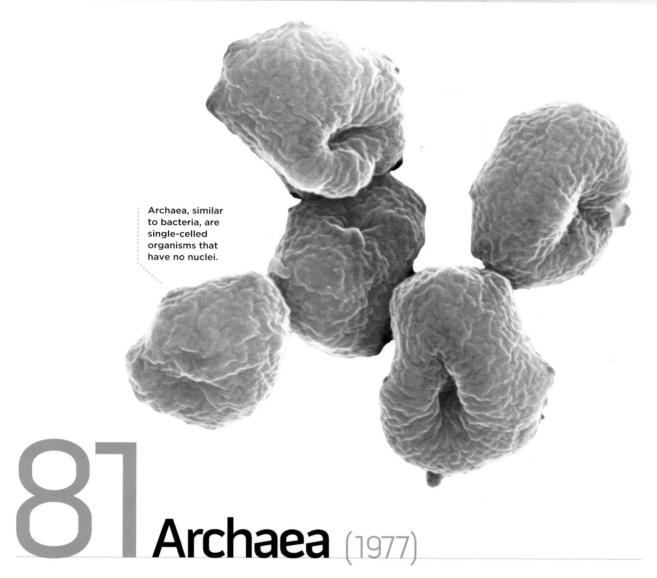

Archaea, similar to bacteria, are single-celled organisms that have no nuclei.

# 81 Archaea (1977)

In 1977 Carl Woese introduced Archaea—his name for a group of single-celled microorganisms that have no nuclei or any other membrane-bound organelles within their cells. Scientists once believed these organisms were simply an anomalous group of bacteria; however, they have a proven independent evolutionary history, and they show many differences from bacteria. In particular, Archaea possess genes and several metabolic pathways and enzymes that are more closely related to those of eukaryotes —organisms whose cells contain complex structures enclosed within membranes.

Archaea were initially considered extremophiles, suitable to live only in the harshest environments, such as hot springs and salt lakes. They have since been discovered in a wide variety of habitats ranging from dirt to deep oceans. In fact, they are especially abundant in the oceans. Archaea in plankton may be one of the most abundant organisms known to humankind.

>> **FACT:** Archaea were first discovered in extremely hot conditions, such as volcanic hot springs.

# 82 Geoengineering

Counteracting the effects of global climate change via large-scale engineering of Earth's environment may sound like the plot of a Hollywood movie, but scientists are taking this idea quite seriously.

Some geoengineering techniques are based on carbon sequestration, including the reduction of greenhouse gases in the atmosphere by capturing carbon dioxide in the air or deliberately depositing iron in the oceans to lock up carbon dioxide in the seabed. Another proposed project is solar radiation management, which is designed to reduce the amount of sunlight hitting Earth. This could be done any number of ways: using pale-colored roofing and paving materials (called cool roofs), using fine seawater spray to whiten clouds and to increase cloud reflectivity, obstructing solar radiation with space-based mirrors or other structures, or cloud seeding.

The use of vertical ocean pipes to mix cooler deep water and warmer surface water is another idea on the horizon. Some proponents—including Bill Gates, chairman of Microsoft—believe this technology could also be used to disrupt hurricanes. Efforts are also underway to counteract climate change by preventing further loss of Arctic sea ice. Many people believe that this project is vital, given the role of Arctic sea ice in maintaining Earth's reflectivity and in keeping methane, a greenhouse gas like carbon dioxide, locked up in permafrost.

Several of these techniques have side effects, and so far no large-scale geoengineering projects are in progress. Some limited tree planting and cool roof projects are underway, and ocean iron fertilization is at an advanced stage of research. Scientists have started field research into sulfur aerosols, which, when shot into the stratosphere by volcanic eruptions, cause the planet to cool temporarily. Reviews of geoengineering techniques emphasize that geoengineering is not a substitute for emission control. Scientists must continue to develop ways to reduce greenhouse emissions in general.

• **87** The concept of what it means to be an **ECOSYSTEM** is crucial to understanding how geoengineering might affect Earth.

• **91** The idea of **CLIMATE CHANGE** as a result of carbon emissions was first put forward in the late 19th century.

One proposal to combat global warming involves blocking sunlight with a huge sun shield.

# 83 Gaia Hypothesis (1970)

The Gaia hypothesis, suggested by James Lovelock (below left), involves the interconnected systems of Earth (below, in cross section).

In 1965 James Lovelock published the first scientific paper suggesting the Gaia hypothesis, a theory that he continued to develop in the late 1960s and early 1970s. Lovelock defined Gaia as "a complex entity involving the Earth's biosphere, atmosphere, oceans, and soil; the totality constituting a feedback or cybernetic system which seeks an optimal physical and chemical environment for life on this planet."

To buttress his theory, Lovelock pointed to the fact that Earth's surface temperature has remained constant even though the energy provided by the sun has increased. And the composition of the atmosphere has remained constant, as has the salinity of the ocean. If extraterrestrial, biological, geological, or other forces change the conditions for life on Earth, life responds to them by modifying and regulating the state of Earth's environment in its favor. In this way Earth, together with all life on the

planet, can be viewed as something like a single organism that maintains equilibrium among its various parts.

**>> FACT:** In Greek mythology Gaia is the goddess of Earth. She gave birth to the sea and the sky.

Buzz Aldrin walks on the moon in 1969.

# 84 Apollo 11 (1969)

On July 20, 1969, humans successfully landed on the moon. Astronaut Neil Armstrong, the first to set foot on the celestial body, uttered his famous phrase, "That's one small step for [a] man, one giant leap for mankind." During their time on the moon, Neil Armstrong and fellow crew member Buzz Aldrin collected 47 pounds of material for analysis.

There were six lunar missions in the Apollo program. Aside from Apollo 11, the one that most people remember is the ill-fated Apollo 13 mission, during which a malfunction on the ship nearly caused the crew to be stranded in space. Apollo 1 never made it off the launchpad; instead, a fire broke out in the command module, and the three astronauts on board were killed. Apollo missions 7 through 10 were never supposed to land on the moon; instead they orbited it and tested the command and lunar modules used by the other missions.

The missions that did make it to the moon conducted experiments on solar wind, soil, meteoroids, seismic activity, and heat flow. These missions were significant not only for their historical importance but also for the data they gathered, which enabled scientists to speculate about the moon's formation and composition.

Yuri Gagarin prepares for launch on Vostok 1.

# 85 Human Spaceflight (1961)

The United States may have been the first nation to put a man on the moon, but it was not the first to put a man into space. That distinction goes to the Soviet Union. On April 12, 1961, Yuri Gagarin made a 108-minute orbital flight at a height of 188 miles as part of the Vostok 1 mission. Although this was considered a significant political victory for the Soviet Union, the United States followed with a manned foray into space a month later. On May 5, 1961, Alan Shepard became the first American to go into space.

Although Gagarin could have controlled his spacecraft manually, it was operated automatically. He was given an envelope with the override code in case of emergency, but psychologists had advised against giving him total control from the beginning of the trip because no one really knew how being in space would affect Gagarin psychologically.

**>> FACT:** Gagarin said that while in orbit he hummed a patriotic song about the motherland.

# 86 Re-creation of Early Atmosphere (1953)

A 1953 experiment in which chemist Stanley Miller combined hydrogen, water, methane, and ammonia provided compelling evidence for how life might have formed on Earth. Miller combined these elements in a flask to simulate Earth's early atmosphere. He sent an electric charge through the flask, which was designed to simulate ultraviolet radiation from the sun. The heat from the charge caused the water in the enclosed microenvironment to be recycled, just as the planet's atmosphere recycles rainwater, oceans, lakes, and other forms of surface water. Miller left the apparatus to operate for a week. When he returned to the flask, he discovered that a scum containing organic compounds, including amino acids, had built up on the surface of the liquid. Around this time, some scientists were speculating that the origins of life could have come from comet dust, which is known to carry amino acids.

Miller was not the first to come up with this theory. In 1922 Russian biochemist Aleksandr Oparin first suggested that Earth's early atmosphere contained hydrogen, methane, water, and ammonia. Oparin believed that life originated in the atmosphere and continued to develop in a prebiotic form in the oceans. His theory was slightly different in that he believed life formed rather spontaneously, whereas Miller acknowledged that the true spark of life had yet to be determined.

>> **FACT:** In 1983 Miller won the Oparin Medal, awarded for important contributions to the study of life's origins.

Dr. Stanley Miller in his laboratory in 1983

# 87 Ecosystem (1935)

However unremarkable it may seem today, the word "ecosystem" represents an insight crucial to the understanding of life on Earth. It was coined in 1935 by British ecologist and botanist Arthur Tansley as a derivation of the phrase "ecological system." Tansley thought the word encapsulated the link between living and nonliving things. Prior to the ecosystem's introduction, animals and plants were believed to interact with each other, but the important roles of soil, water, and air as components of the environment had been ignored or overlooked. With Tansley's ecosystem, scientists acknowledge that the whole may very well be greater than the sum of its parts.

**>> FACT:** In 2008 Ecuador became the first country to recognize ecosystem rights in its constitution.

Red mangrove roots penetrate the underwater soil in Florida.

The big bang emitted energy in such immense amounts that scientists continue to measure it today.

# 88 Big Bang Theory (1931)

According to the big bang theory, the universe was initially in an extremely hot, dense state that expanded rapidly; it has since cooled by expanding to its present diluted state, and it continues to expand. Belgian physicist Georges Lemaître first proposed this theory in 1931. The theory is often put forth as an explanation of how the universe came into being. Rather, it describes the general evolution of the universe since it came into being.

The kernel of this theory lies in observations of spiral galaxies in 1912, which American astronomer Vesto Slipher showed to be receding from Earth. In 1929 Edwin Hubble discovered that all remote galaxies and clusters are moving directly away from Earth's vantage point. Two years earlier, in 1927, Lemaître had proposed that the universe was expanding in order to take into account some astronomical observations. In 1931 he took this a step further by suggesting that the universe began as the expansion of a single point— bringing time and space into existence.

01

03

03

02

# 89 Terraforming

Anyone who has read Kim Stanley Robinson's Mars trilogy knows all about terraforming, a term coined by Jack Williamson in the science-fiction story "Collision Orbit" in 1942. But the idea is not just in the realm of science fiction. In fact, for several reasons, Mars is a likely candidate for terraforming—modifying the planet's atmosphere so that it is habitable for humans.

Recently NASA probes discovered hints that water might have flowed on Mars. This is an indicator that life may have existed in the planet's distant past. The water might still be there, trapped in ice caps at the poles, which means life could exist there again—provided scientists can transform the currently cold, dry Martian atmosphere into one like Earth's. In addition to water, Mars holds other elements that are needed for life to exist, including carbon, oxygen, and nitrogen.

So how would scientists manage to make Mars suitable for humans? And more important, if they figure out a way to do it, should they do so? The first question might be easier to answer than the second. Three proposed terraforming methods include

BREAKT

## 01 Not in Our Lifetime
Terraforming Mars will be a thousand-year project, starting with a number of exploratory missions that gradually build an infrastructure.

## 02 Pioneer Village
Each new mission will involve the construction of more habitation modules, extending the stay time and expanding the living space.

## 03 Global Warming
Space mirrors will melt Mars's ice, and engineers will design factories to spew greenhouse gases to make the atmosphere trap warmth.

## 04 Life Under Domes
Geodesic domes will provide climate-controlled living spaces, first for plants and later for humans.

## 05 Power to the Planet
Nuclear power and wind turbines are two current technologies that could be carried to Mars.

## 06 Don't Forget Your Mask
Even after a millennium of development, Mars's atmosphere will not provide enough oxygen for humans to breathe without equipment.

---

installing large mirrors in orbit around Mars to reflect sunlight and heat the planet's surface; constructing greenhouse gas–producing "factories" to trap solar radiation; and smashing asteroids laden with ammonia into the planet to create a greenhouse-like effect.

Regardless of the method used, this is not a process that could be completed in decades—and maybe not even in centuries. Some scientists believe that such a project could take millennia to become a reality. In addition, although the technological capacities to pull off a project of this magnitude are potentially within grasp, the economic capacity is not.

• **84** The accomplishment of the **APOLLO 11** mission—landing humans on the moon—is a direct precursor to landing humans on other planets.
• **85 HUMAN SPACEFLIGHT** makes the idea of terraforming possible, at least in theory.
• **87 & 93 ECOSYSTEMS** and the roles of **OXYGEN AND CARBON DIOXIDE** in Earth-based life are crucial to contemplating terraforming.

HROUGH

# 90 Continental Drift (1915)

In geology, as in other areas of science, some theories are accepted readily while others are not. The latter was the case with continental drift and plate tectonics, which German scientist Alfred Wegener first introduced in 1915 in his book *The Origins of Continents and Oceans*. In his studies Wegener realized that identical fossils of plants and animals were found on opposite sides of the Atlantic Ocean. The common understanding at that time was that land bridges had connected the continents. Wegener, however, noticed the ways in which the continents appeared to have once fit together like puzzle pieces; he speculated that they must have formed a giant landmass, which he called Pangaea (a Greek term meaning "all the earth").

At first the scientific community heaped scorn on Wegener for his theory. As geologists continued to learn more about Earth's composition, however, Wegener's theory of continental drift did not look so silly anymore. The study of plate tectonics has shown that the continents do indeed "float" on a layer of underlying rock.

An illustration of Earth during the Cretaceous period, 100 million years ago

A computer simulation shows global patterns in climate.

# 91 Climate Change (1896)

In 1896 Swedish scientist Svante Arrhenius calculated that the temperature of Earth would increase as the amount of carbon dioxide in the atmosphere increased. Arrhenius did not know it, but he had laid the foundation for a theory that would engender fierce debate a century later. He attributed this increase in carbon dioxide directly to human activity—namely, the burning of fossil fuels—and he calculated that every time the increase in carbon dioxide levels doubled, the temperature would go up by 5°C.

Although Arrhenius was the first to publish this idea formally and to support it with scientific evidence, the idea that humans could have a significant and potentially detrimental effect on the planet was not new. The ancient Greeks, for example, speculated on the relationship between rainfall and the cutting down of a forest, and early farmers were well acquainted with the practice of slash-and-burn agriculture. But could humans really affect things on such a global scale? Although many in Arrhenius's time disagreed with him, the answer, he affirmed, was a resounding yes.

As temperatures in the North Atlantic region warmed by the 1930s, most people were still reluctant to attribute the cause to humans. English engineer G. S. Callendar, however, used improved techniques to refine Arrhenius's calculations in 1938 and sounded the alarm that greenhouse global warming was indeed taking place and should not be ignored. Detailed measurements in the 1960s showed that global warming was not a theory but a fact.

>>**FACT**: Global temperature has been measured since 1880, and the 20 warmest years have occurred since 1981.

# 92 Age of Earth (1830)

When people first started to speculate about the age of Earth, they looked to the Bible for an answer. Based on the account given in the Book of Genesis, Earth was believed to be about 6,000 years old. This remained the prevailing belief until the 19th century, when the field of geology was becoming more accepted as a valid branch of science. Geologists were exploring the planet and making some astounding discoveries: that Earth's strata showed that the land and the oceans and seas had literally traded places at times; that the surface of Earth could be transformed by geologic events like earthquakes and volcanoes; and that forces such as water erosion could change the planet as well, just more slowly. The age of Earth clearly needed to be revisited in light of such observations.

Scientists like James Ussher and Baron Georges Cuvier attempted to use empiricism to reason their way to an answer. The problem was that they still based their reasoning on a biblical framework. In 1830 geologists Charles Lyell and James Hutton were among the first to take a different tack. Rather than attributing evidence for major geological changes to onetime cataclysmic events in the Bible (like Noah's flood), Lyell and Hutton reasoned that changes on the face of Earth were the result of naturally occurring processes that had been present throughout the history of the planet. Thus the age of Earth had to be greater than 6,000 years. Scientists today estimate Earth's age to be around 4.5 billion years.

Scientists have used information about fossils, such as this ammonite, to estimate Earth's age at 4.5 billion years.

>> **FACT:** The oldest rocks found on Earth are the Acasta Gneisses (3.96 billion years) in Canada.

Joseph Priestley used a mouse, a jar, and some mint to discover the lifesaving link between the gases that people and plants take in and release.

# 93 Oxygen and Carbon Dioxide (1774)

In 1774 English chemist and theologian Joseph Priestley conducted a famous experiment in which he proved that plants take in a gas that animals give off (carbon dioxide) and that plants give off a gas that animals take in (oxygen).

Priestley took a mouse and placed it in an enclosed glass container until the mouse collapsed. When he put a plant in the container with the mouse, the mouse survived, thereby proving that the plant was producing something that enabled the mouse to live. Priestley did not know that the substances were oxygen and carbon dioxide, per se, but, in fact, these were the substances that he had discovered.

# 94 Discovery of Earth-like Planets

To date, over 600 alien planets have been discovered. Could any of them be similar to Earth? Leaders of projects like SETI (Search for Extraterrestrial Intelligence) have tried to determine this by scanning the heavens with radio telescopes, but this technique has its limitations. (What if life on other planets has not developed the means to transmit radio signals yet?) Scientists from NASA think they have come up with a way to find Earth-like planets based on color alone. When we look at planets through a set of filters—red, green, and blue—Earth stands out among the others, in that it appears much bluer. This is because Earth's atmosphere is low in infrared-absorbing gases like methane and ammonia, compared to planets like Jupiter and Saturn.

Various investigators are searching for the presence of extrasolar planets (exoplanets), particularly those in orbits where temperatures would allow for liquid water, and for clues to life-supporting chemicals in planetary atmospheres. The Keck Interferometer, for example, uses a technique known as interferometry to combine the light of the world's largest optical telescopes in order to study dust clouds around stars where Earth-like planets may be forming. In March 2006 the first mission to look for Earth-like planets in the Milky Way was launched, and the Kepler spacecraft, with its giant telescope, was sent into space.

SIM PlanetQuest, which will follow Kepler, will measure the distances and positions of stars with unprecedented accuracy, allowing scientists to locate planets in the zones around nearby stars that look like they could be conducive to life.

So far, astronomers have primarily discovered huge planets that likely do not contain life. However, given the vastness of the universe and the billions of planets that the Milky Way contains, it is surely only a matter of time before planets like Earth are discovered—perhaps ones that support life-forms. Finally, the holy grail of astronomy would be achieved: indisputable proof that we humans are not alone.

• **78** The **HUBBLE TELESCOPE** is one of the main tools for the discovery of extrasolar planets.

• **97** The **TELESCOPE** is an important precursor to the tools used today to scan the galaxy for planets that might have Earth-like qualities.

• **98** The **COPERNICAN SYSTEM** changed humanity's place in the cosmos; without it scientists would not be searching for other planets like Earth.

Astronomers hypothesize that conditions that support life may exist on planets outside our solar system. An artist's version depicts the view from orbit of an extrasolar planet and, beyond it, two moons and its distant sun.

A fossil of
a tiger
shark tooth

# 95 Fossils (1669)

Nicholas Steno is noted for his theory of fossil formation in rock layers, which he put forth in his 1669 work *Prodromus*. Steno's theory was crucial to the development of modern geology, and his ideas are still very much in use today. When examining a shark's teeth, the Danish pioneer in geology was struck by how similar they looked to objects known as tongue stones that were often found in rocks. He correctly assumed that the tongue stones were not stones after all, but were actual sharks' teeth that had somehow become embedded in rock. But how? He surmised that what was solid rock had once been liquid. When an item such as a shark's tooth is deposited in the liquid and that liquid hardens, it takes the shape of the item and produces a fossil.

>> **FACT**: Fossilized single-celled organisms that are 3.4 billion years old are the oldest known fossils on Earth.

# 96 Kepler's Laws of Planetary Motion (1609)

German astronomer and mathematician Johannes Kepler formulated his first two laws of planetary motion in 1609 and his third law around 1619. His work drew on the observations of his mentor, Tycho Brahe. The laws are as follows:

(1) Each planet moves in an elliptical orbit, with the sun at one focus of the ellipse.

(2) A line from the sun to each planet sweeps out equal areas in equal time. This law implies that a planet will move faster when it is closer to the sun.

(3) The square of a planet's orbital period is proportional to the cube of the distance from the sun. In other words, the time it takes a planet to orbit the sun is related to its distance from the sun.

Although these laws seemed to be true based on observation, it would be nearly a century before anyone could prove that they were correct. Newton's law of universal gravitation proved that Kepler's laws actually do describe the motion of the planets in orbit. Today, scientists know that Kepler's laws apply only approximately to motions in the solar system; however, they are close enough for most purposes.

>> **FACT:** NASA named a 2009 mission to find habitable Earth-size planets after Johannes Kepler.

Orbits of our solar system's outer planets and dwarf planets closely conform to laws articulated 400 years ago.

# 97 Telescope (1609)

Although Galileo was not the first to invent the telescope, he made significant improvements to its design in 1609 and was undoubtedly the first person to make such amazing use of it. His telescope was unique in that it produced upright images. Although the images it produced were blurry and distorted, the instrument was still good enough for Galileo to explore the sky. He was the first to document the phases of Venus, the craters on the moon, and the four largest moons orbiting Jupiter, for example. Galileo's observations confirmed for him that Earth was definitely not the center of the solar system, as the Aristotelian view of cosmology claimed; rather, Galileo began to see that the Copernican view was correct: The sun was the center of the solar system.

A telescope at Arizona's Kitt Peak National Observatory

# 98 Copernican System (1543)

Published in 1543 by Polish astronomer Nicolaus Copernicus, the Copernican system, which places the sun at the center of the solar system, overturned preconceived notions not only about the cosmos, but also about humankind's place within it. Prior to Copernicus, the prevailing theory was the Ptolemaic system. In this view of the universe, Earth was at the center, stars were embedded in a large celestial sphere, and the other planets were in smaller spheres in between. Ptolemy's system included epicycles—smaller circles in which the planets moved while they orbited Earth—in order to explain the apparent backward motion that the planets sometimes exhibited.

Copernicus described his system as follows: The motions of heavenly bodies are uniform, eternal, and circular or made up of several circles, or epicycles; the center of the universe is near the sun; and around the sun, in order, are Mercury, Venus, Earth and its moon, Mars, Jupiter, Saturn, and the fixed stars. It would take 200 years for the Copernican model to replace the Ptolemaic model.

**Diagrams of the systems of Ptolemy and Tycho Brahe surround an illustration of Copernicus's sun-centered vision.**

Solid rocket boosters carry the space shuttle *Discovery* skyward.

# 99 Rocket (1232)

Who made the first rocket? The answer depends on the definition of the term. Various cultures have experimented with rocket-like devices, and the concept dates as far back in recorded history as 400 B.C., when a Greek inventor heated a hollow ball full of water and escaping steam propelled it.

Sources indicate that the first documented use of rockets was in China. The Chinese had invented gunpowder made of charcoal dust, sulfur, and saltpeter (potassium nitrate); they discovered that when this substance was packed into hollow bamboo tubes and thrown into a fire, some of the tubes exploded. They attached arrows to such tubes to stabilize their paths, and thus the first military rockets were born. The rockets' first recorded use was in the battle of Kai-fung-fu, in 1232, when "fire arrows" helped villagers defend themselves against invading Mongols.